D1501307

THE BODY FACTORY

THE BODY FACTORY

FROM THE FIRST PROSTHETICS TO THE AUGMENTED HUMAN

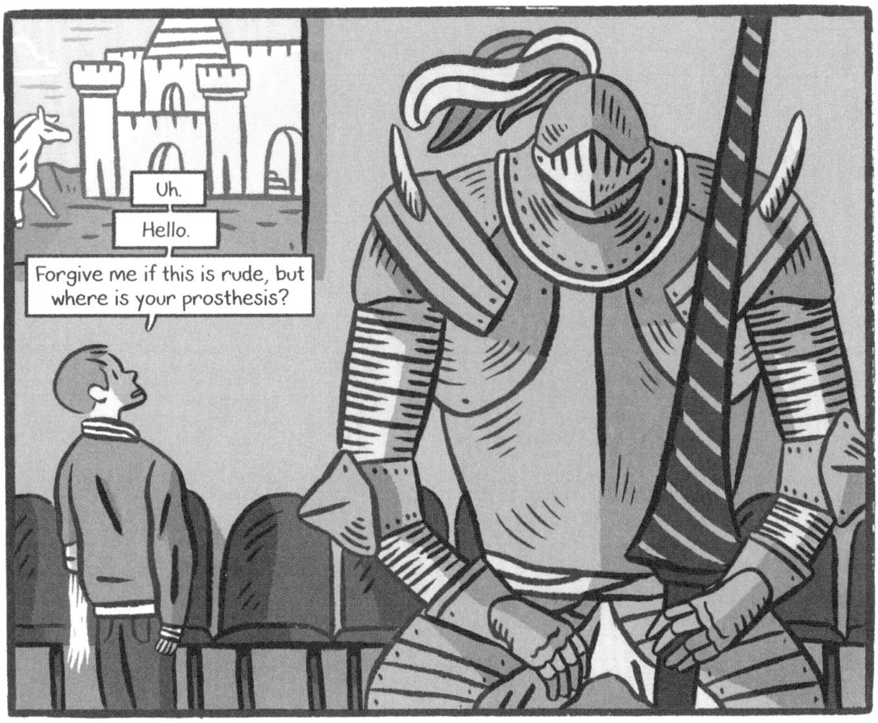
Uh.
Hello.
Forgive me if this is rude, but where is your prosthesis?

STORY AND ART

HÉLOÏSE CHOCHOIS

GRAPHIC MUNDI

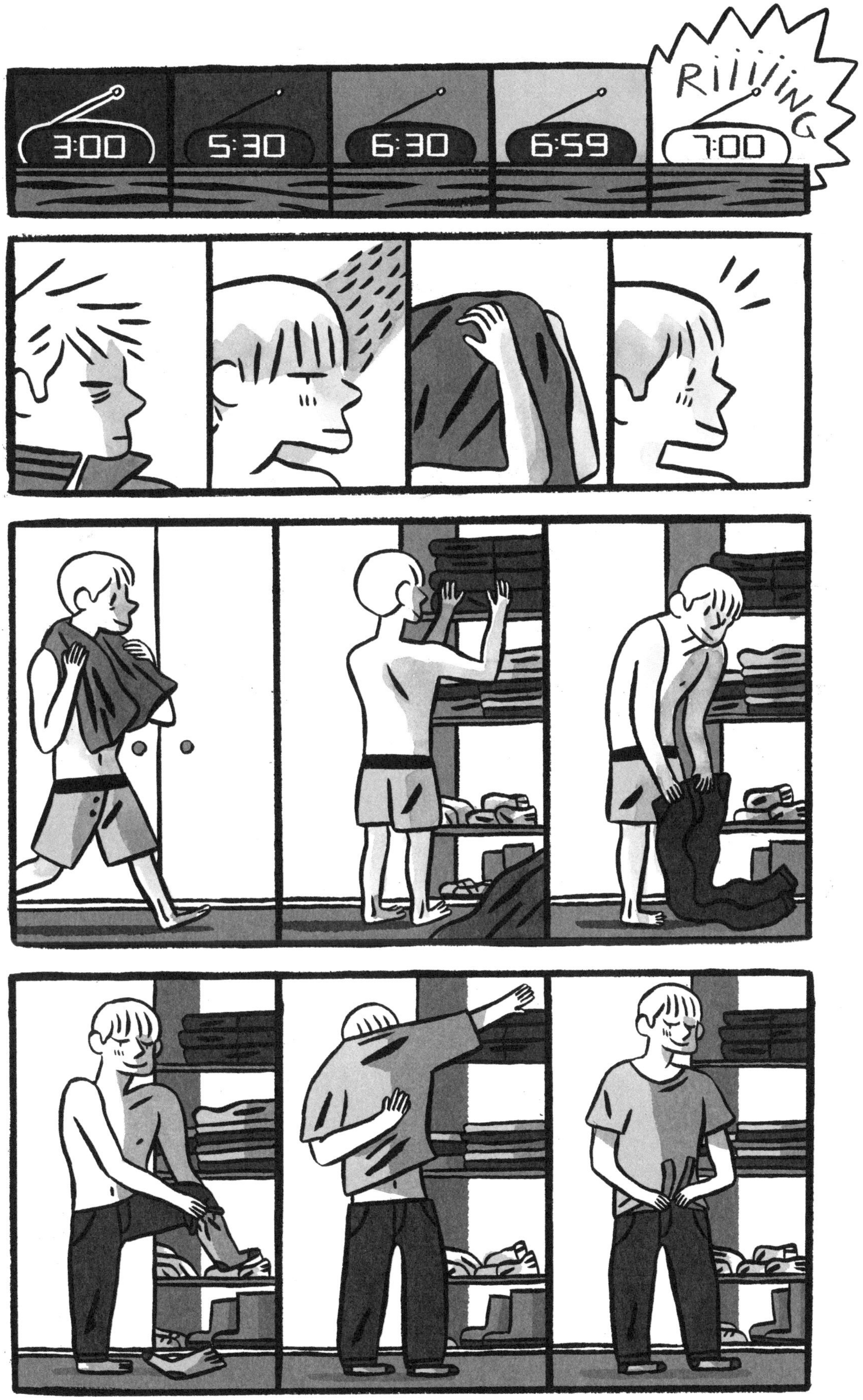
3:00
5:30
6:30
6:59
7:00
RIIIIING

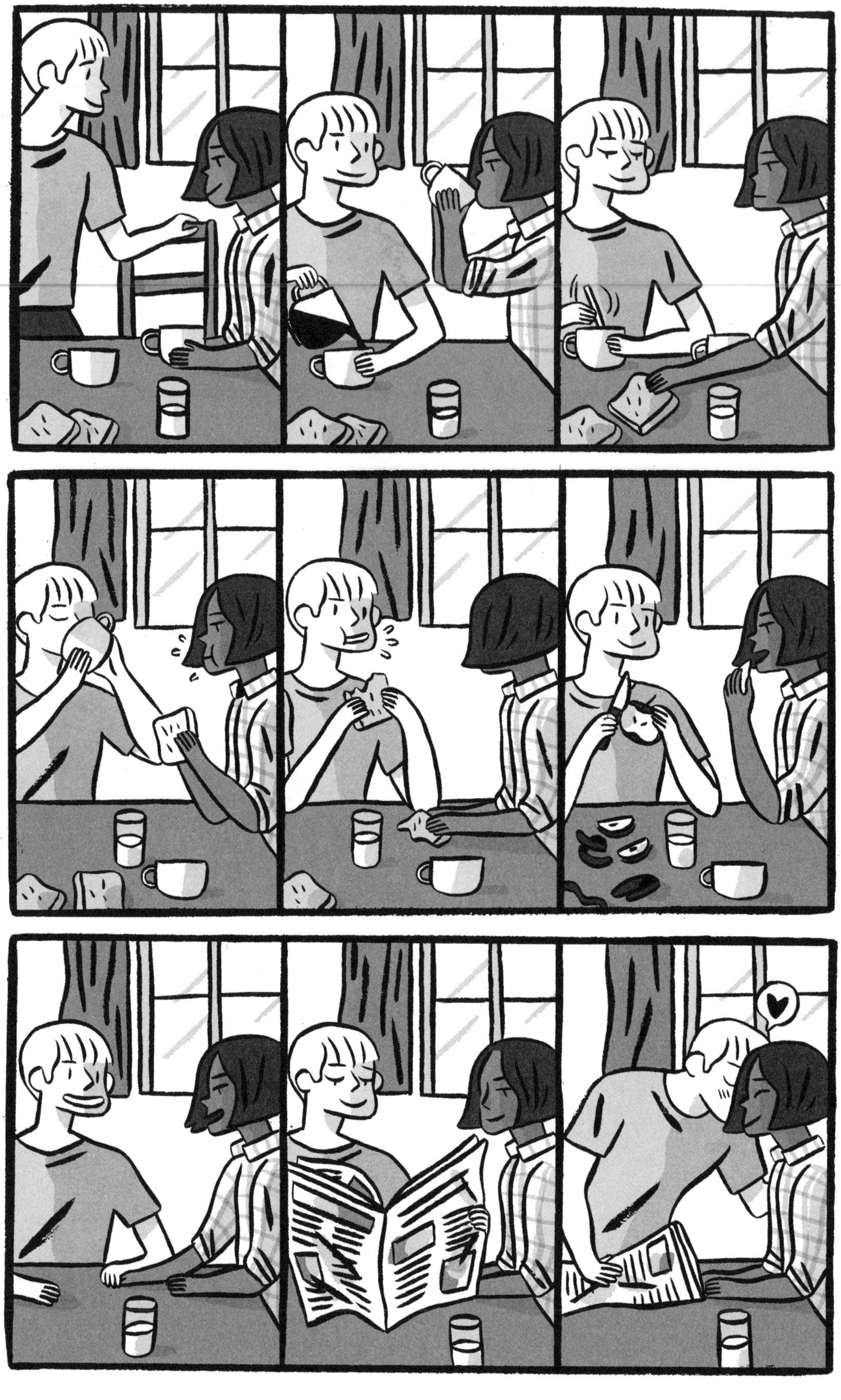

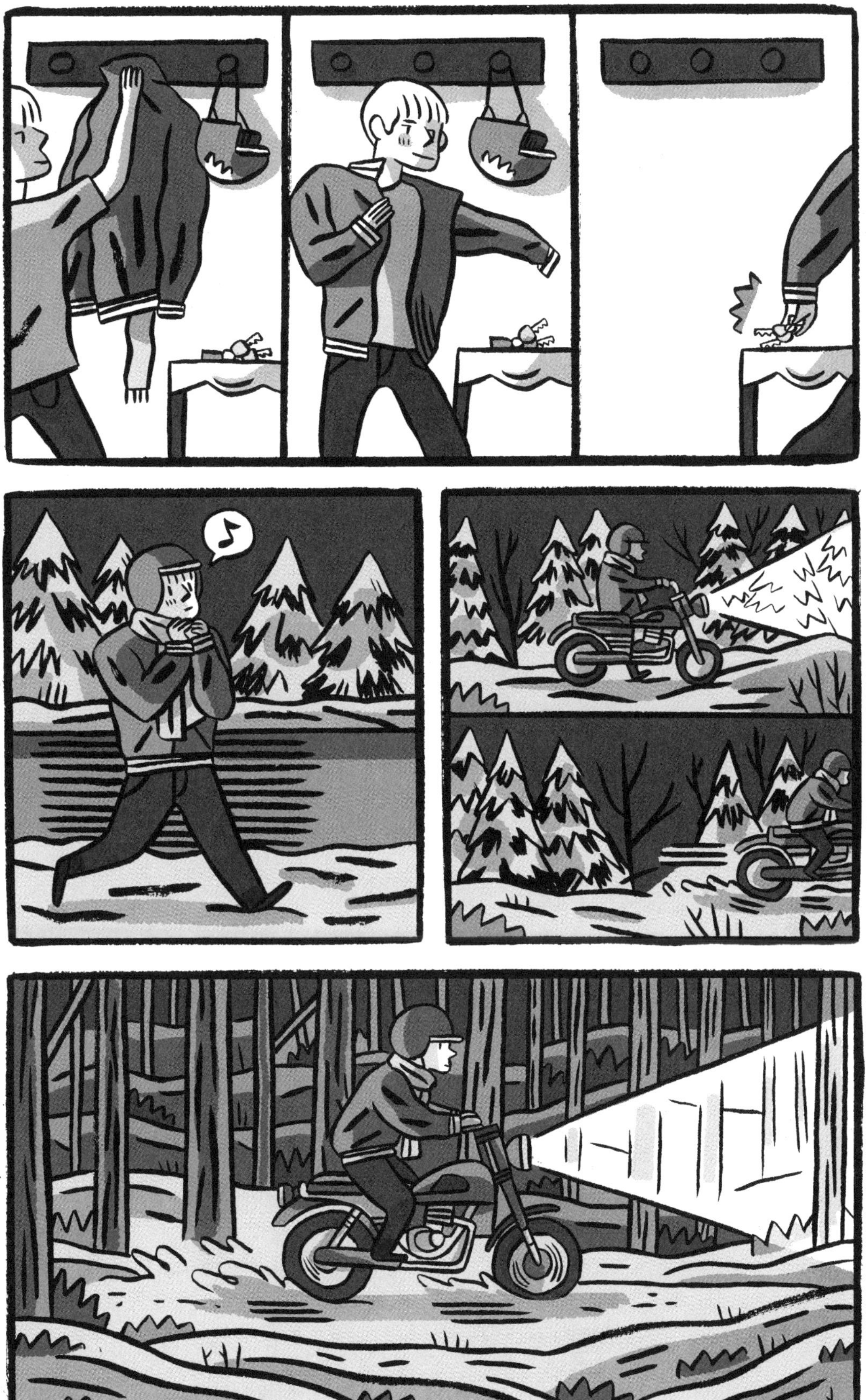

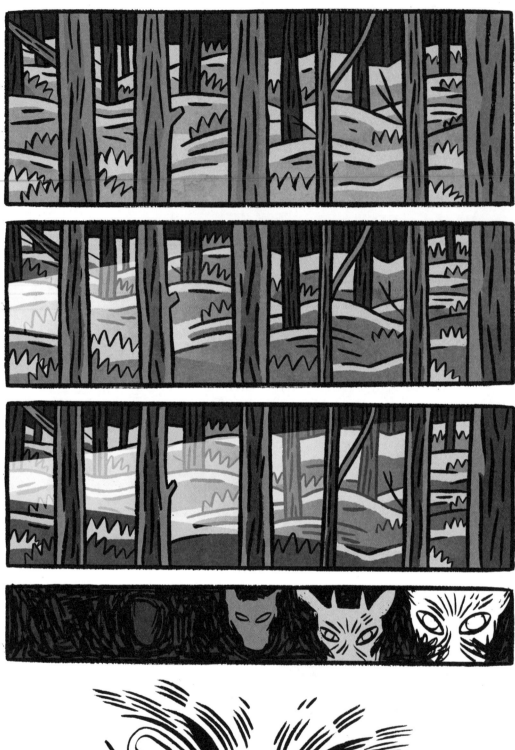

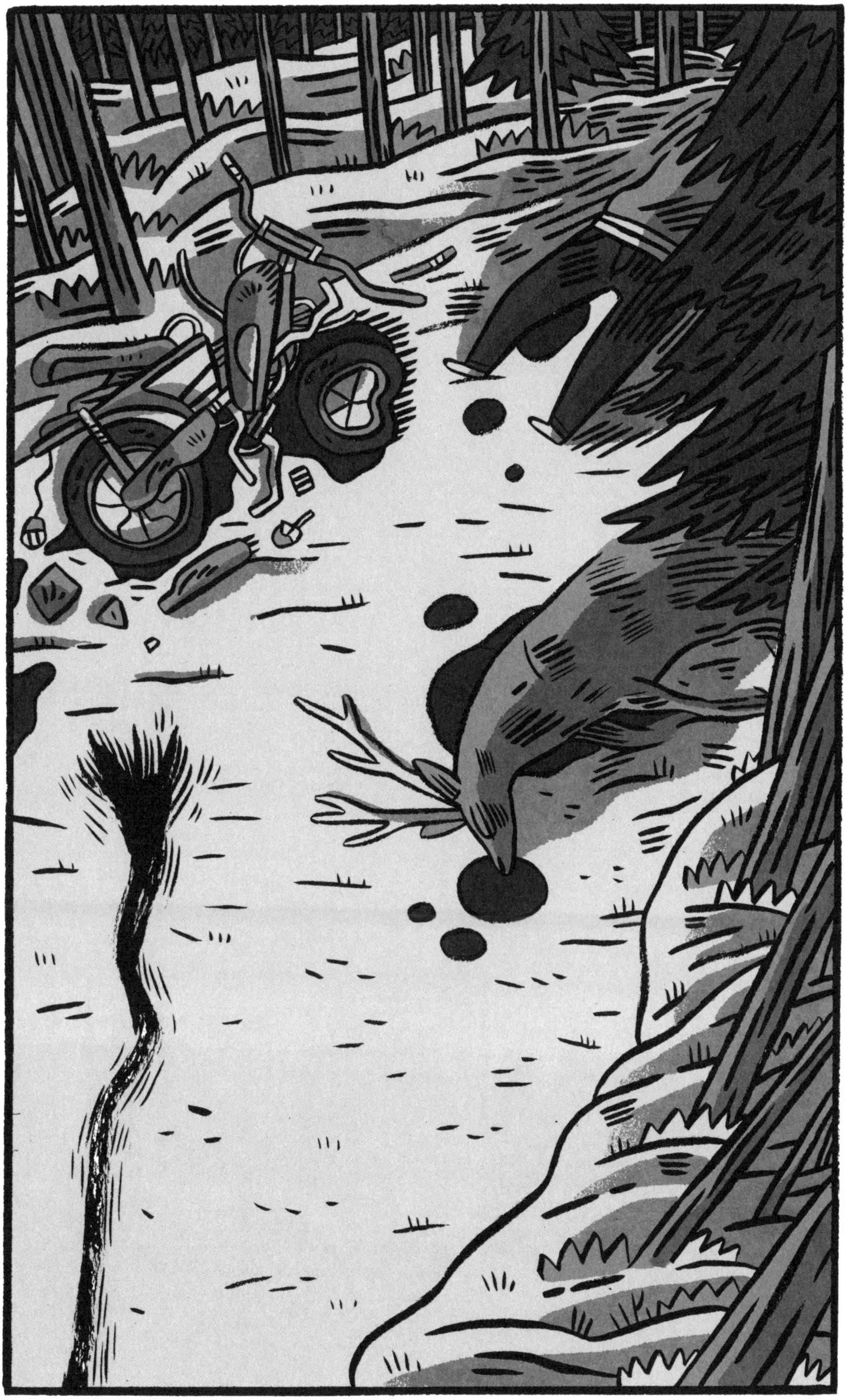

CHAPTER 1
Amputation

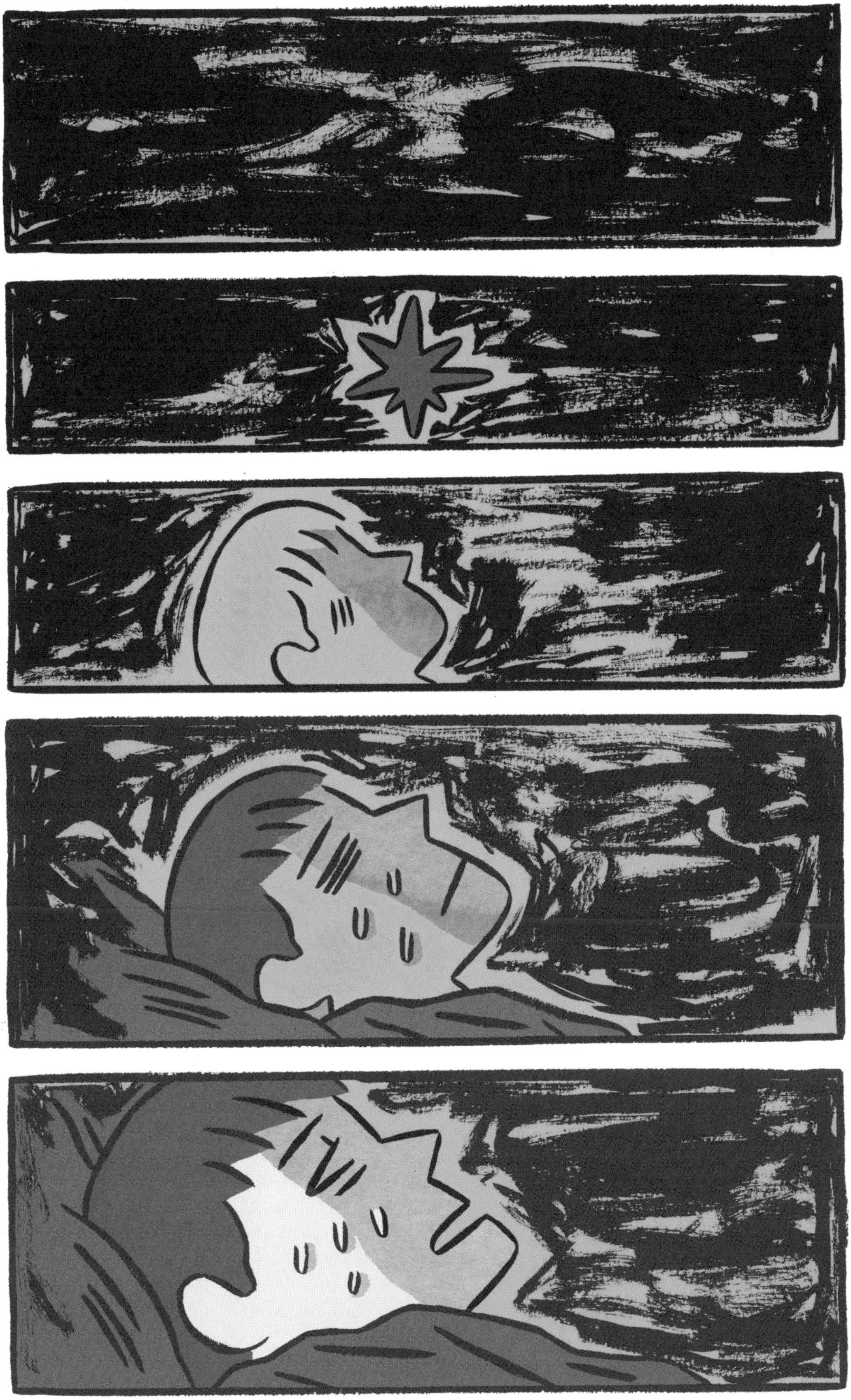

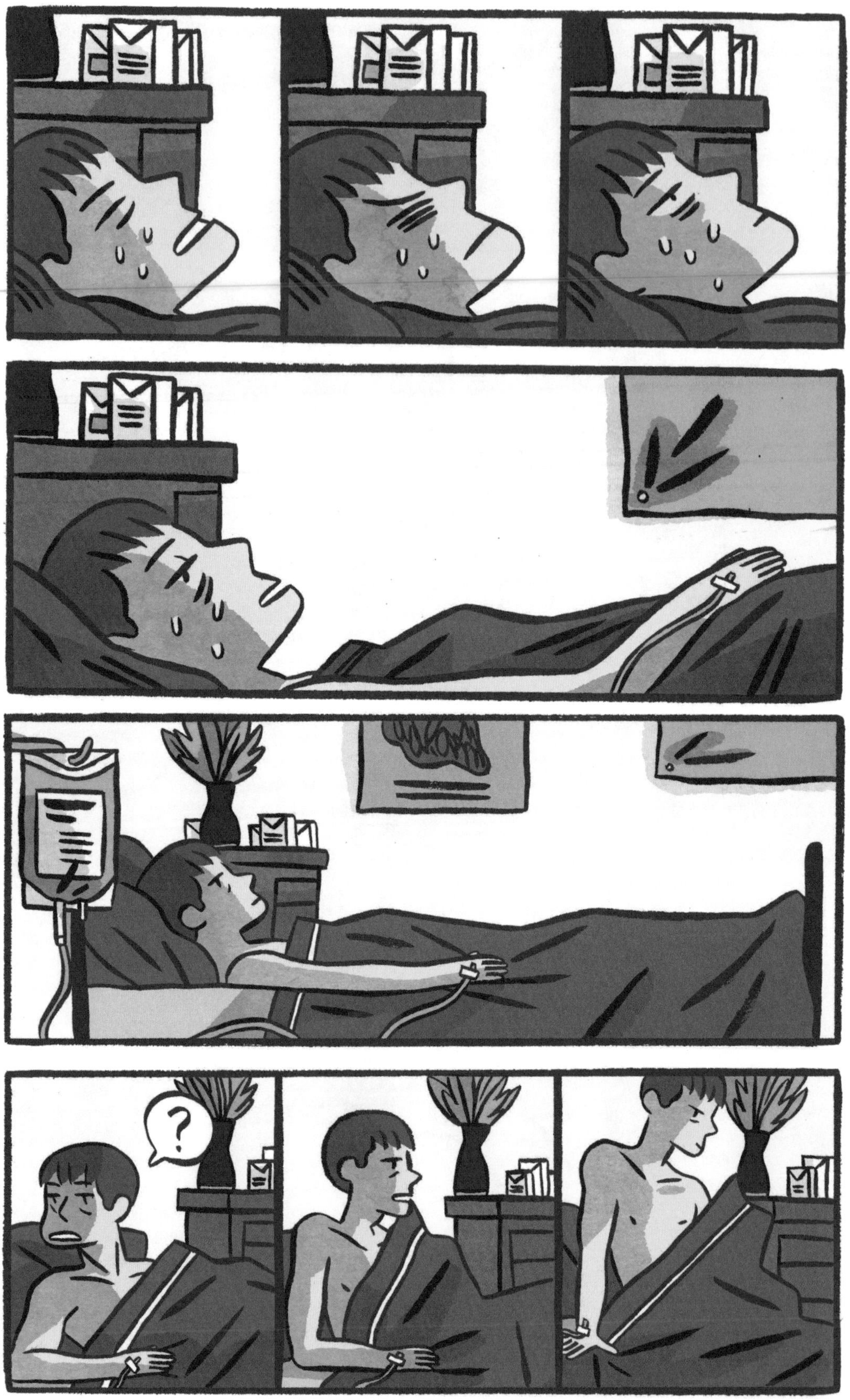
?

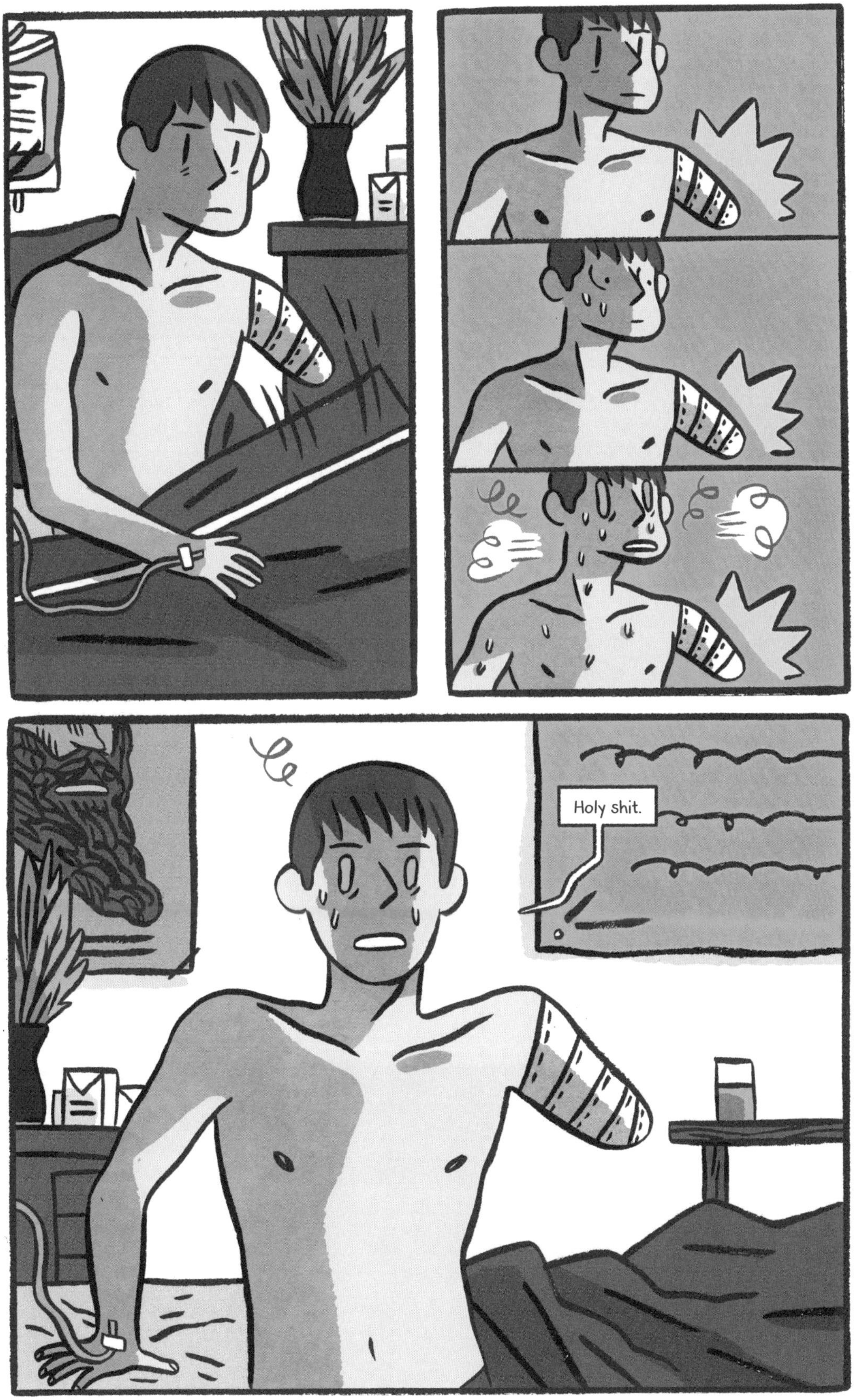
Holy shit.

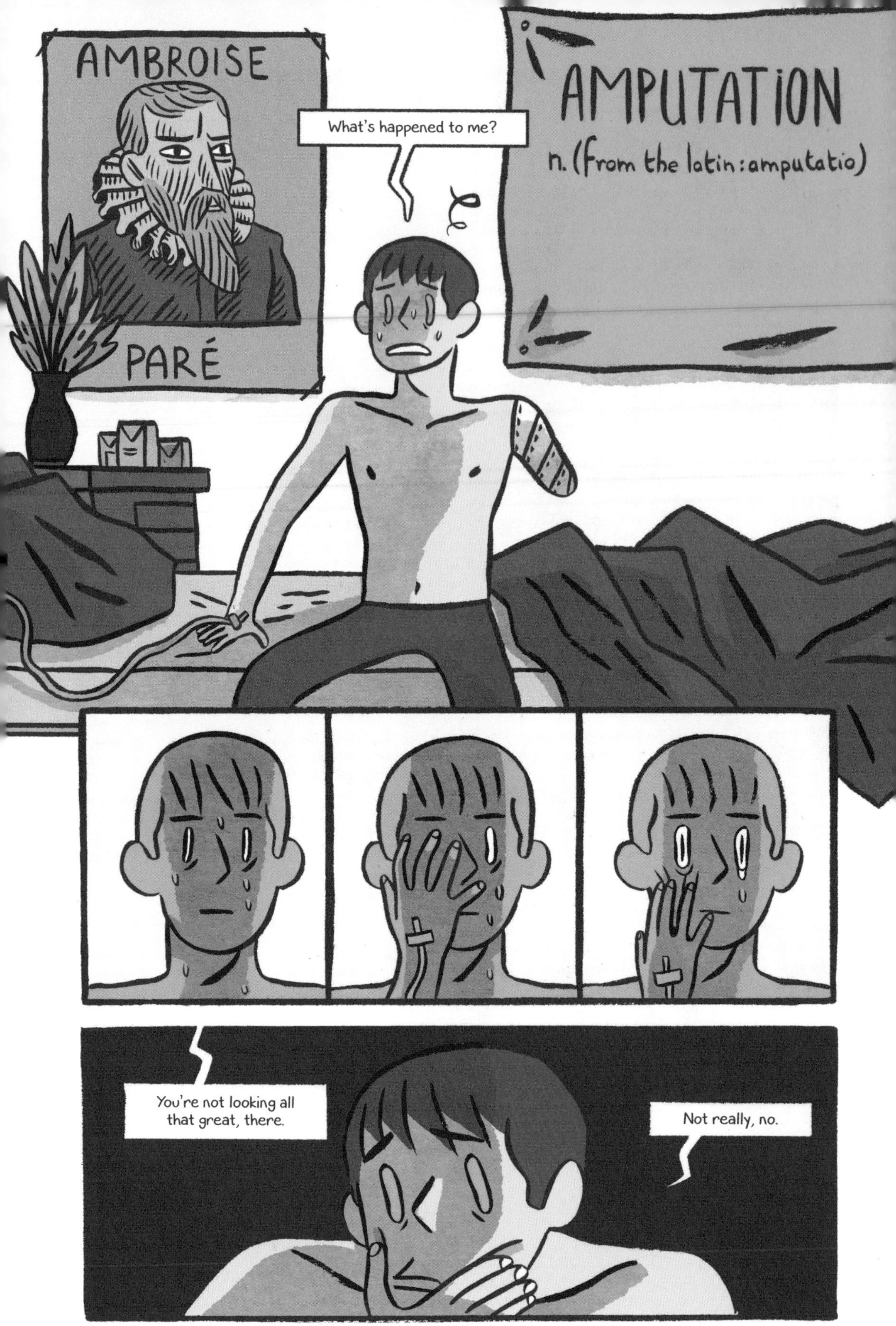
AMBROISE
PARÉ
What's happened to me?
AMPUTATION
n. (from the latin: amputatio)
You're not looking all that great, there.
Not really, no.

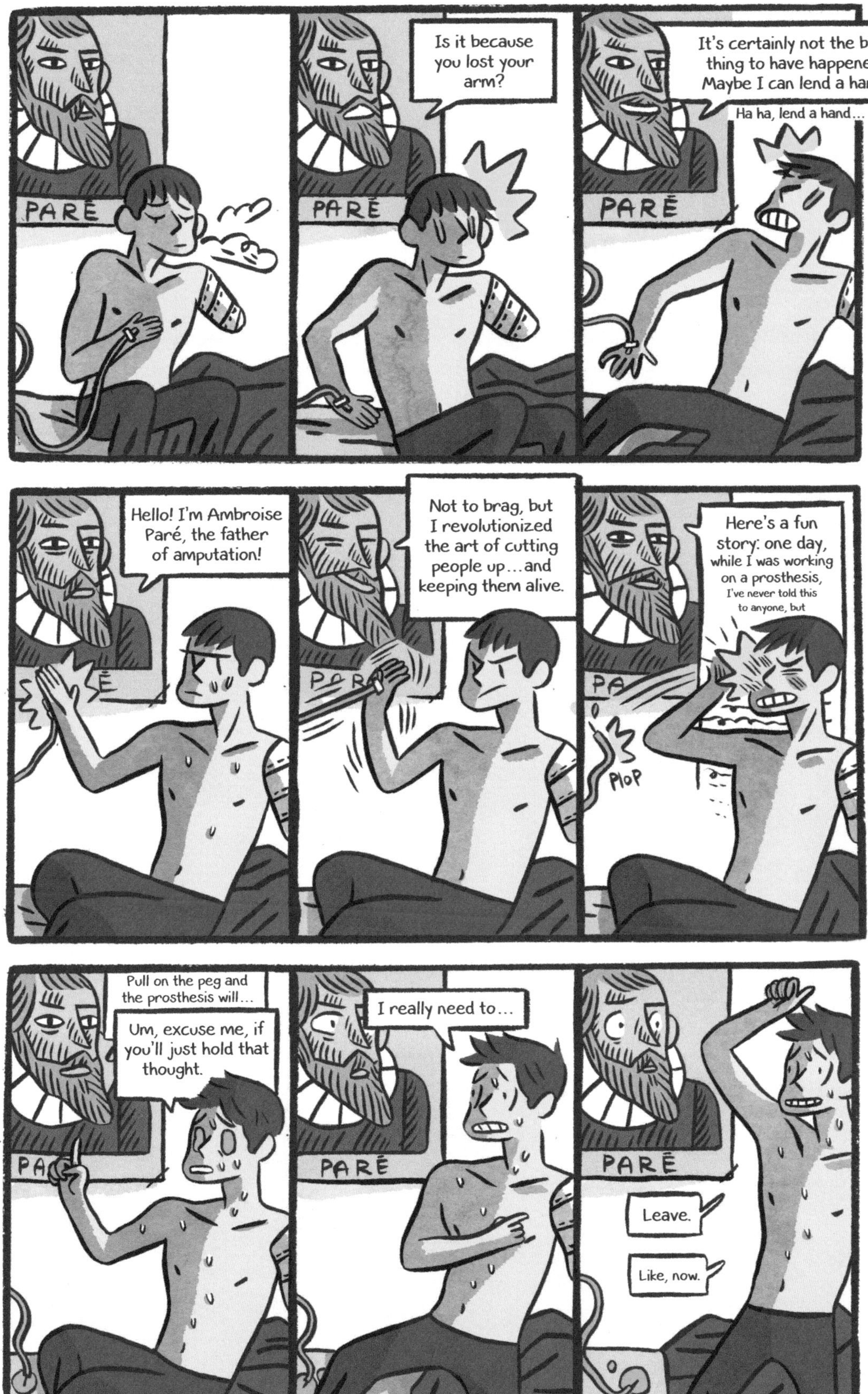
PARÉ
Is it because you lost your arm?
PARÉ
It's certainly not the best thing to have happened. Maybe I can lend a hand!
Ha ha, lend a hand...
PARÉ
Hello! I'm Ambroise Paré, the father of amputation!
Not to brag, but I revolutionized the art of cutting people up... and keeping them alive.
Here's a fun story: one day, while I was working on a prosthesis,
I've never told this to anyone, but
PLOP
Pull on the peg and the prosthesis will...
Um, excuse me, if you'll just hold that thought.
I really need to...
PARÉ
PARÉ
Leave.
Like, now.

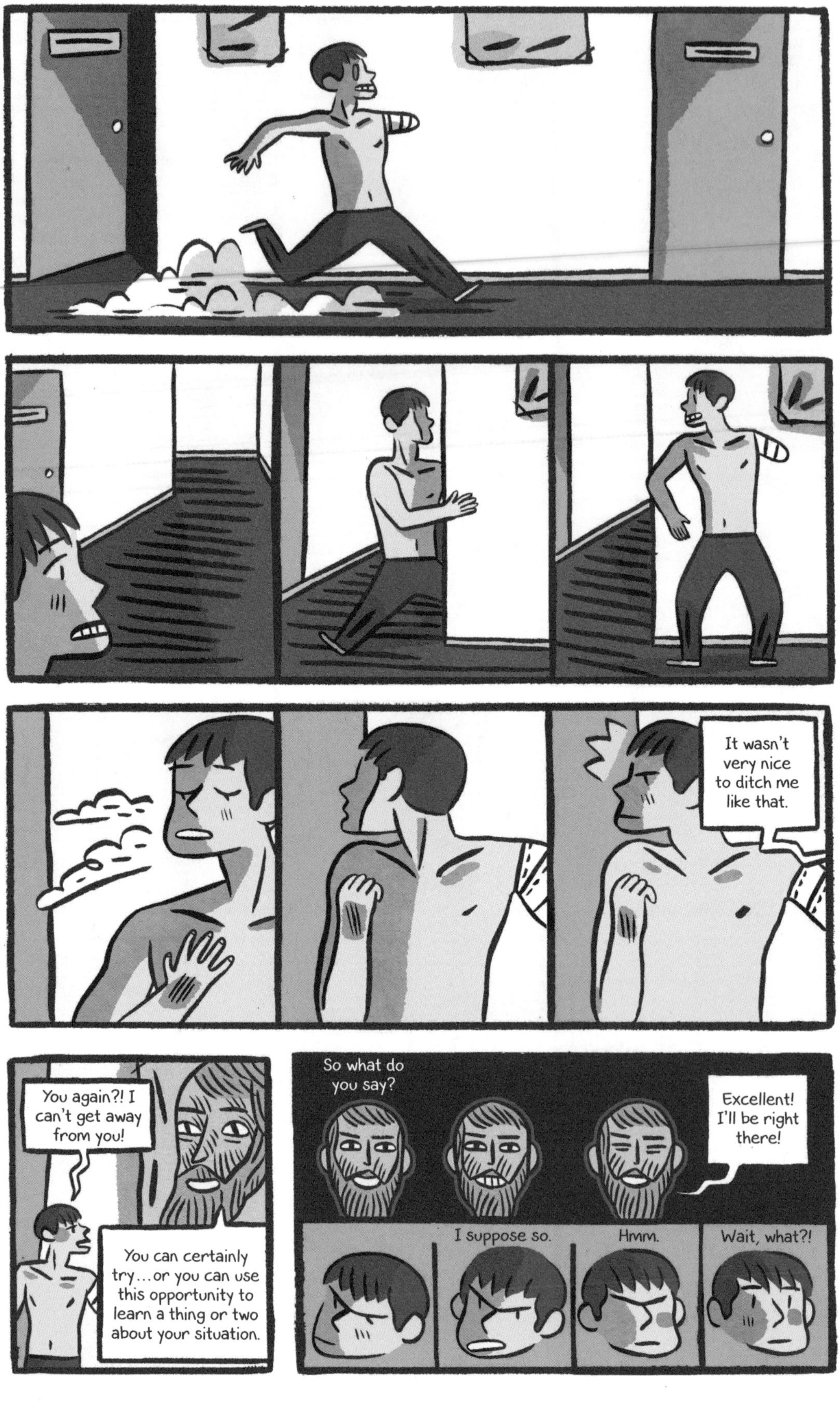
It wasn't very nice to ditch me like that.
You again?! I can't get away from you!
You can certainly try...or you can use this opportunity to learn a thing or two about your situation.
So what do you say?
I suppose so.
Hmm.
Excellent! I'll be right there!
Wait, what?!

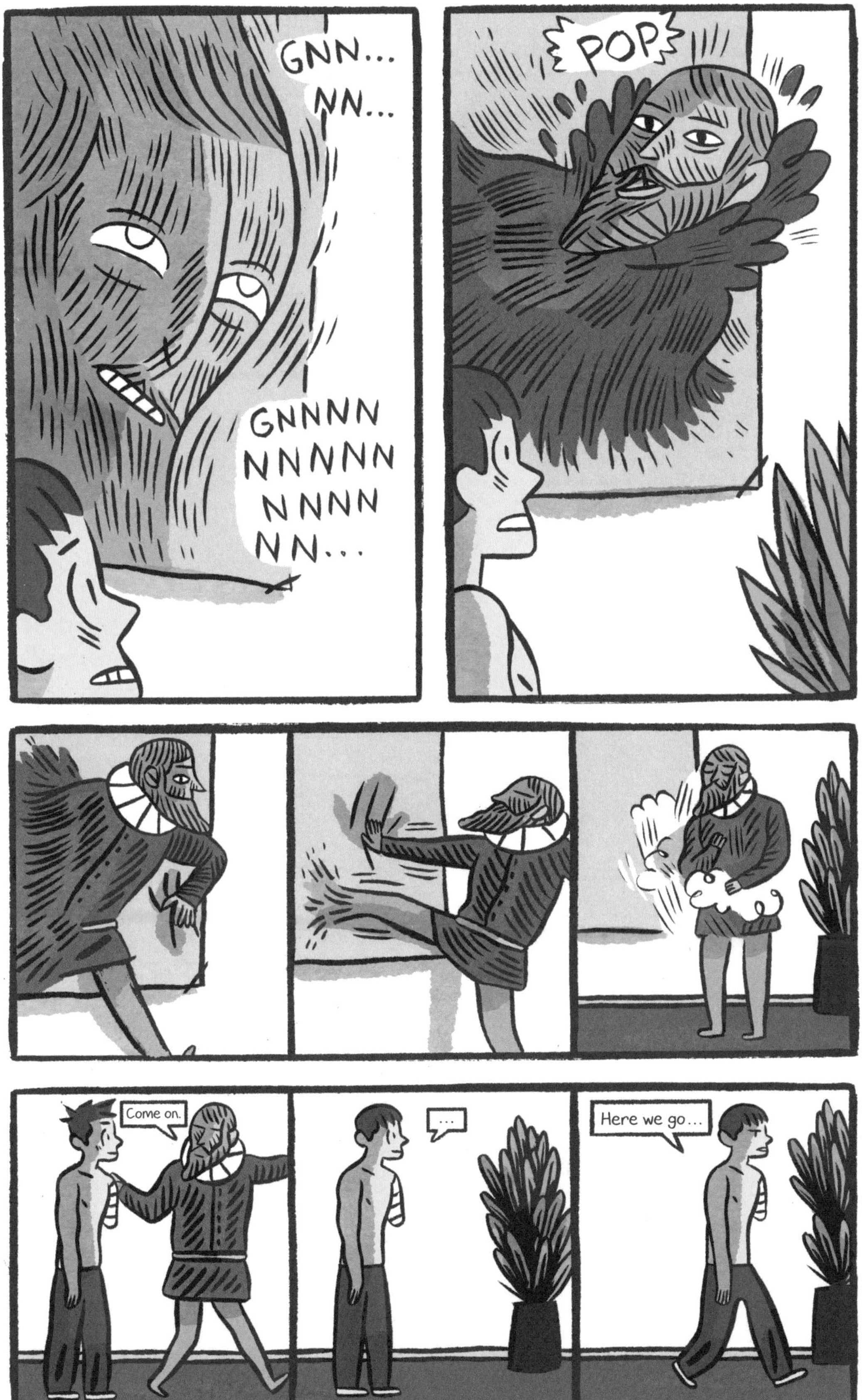
GNN...
NN...
GNNNN
NNNNN
NNNN
NN...
POP
Come on.
...
Here we go...

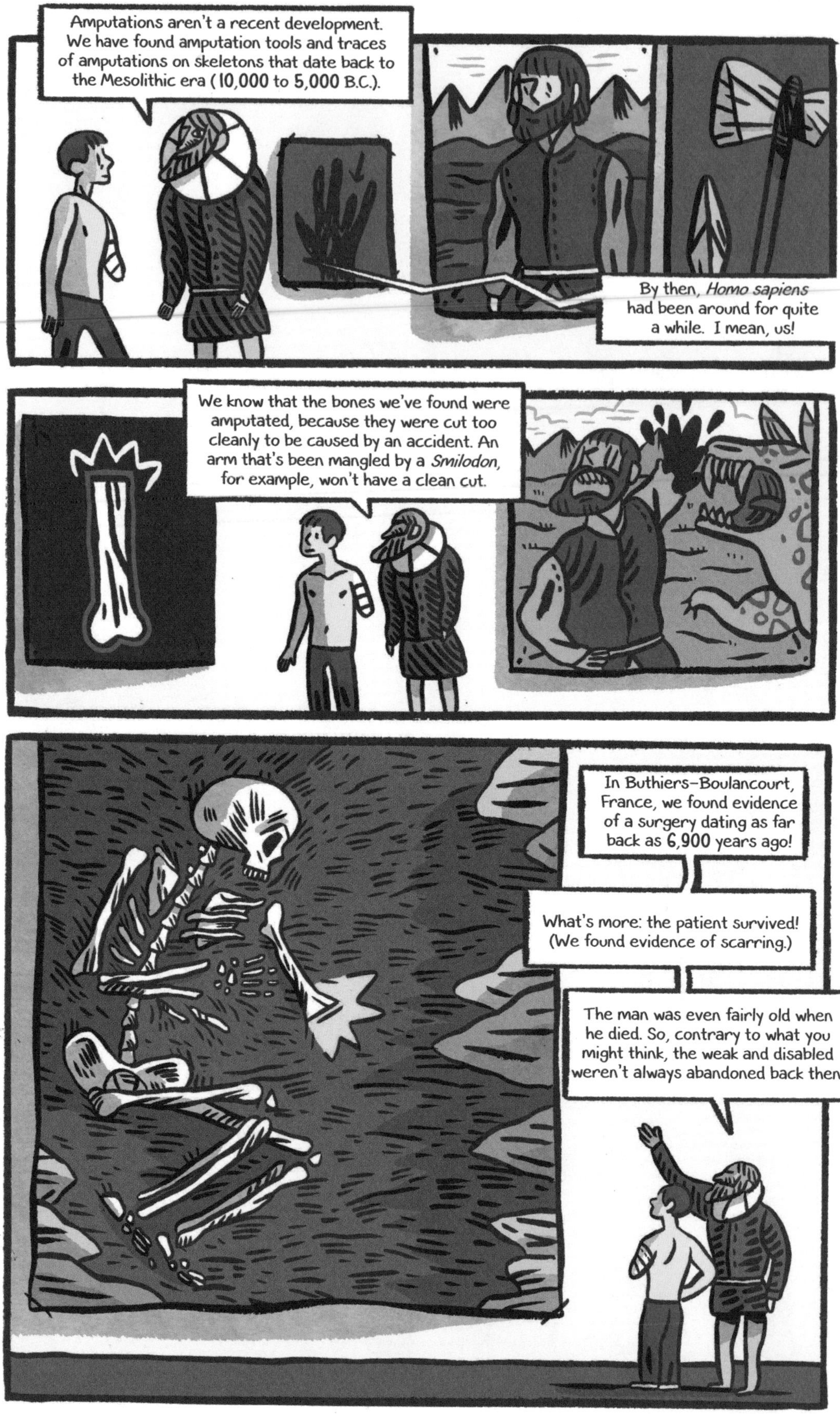

Amputations aren't a recent development. We have found amputation tools and traces of amputations on skeletons that date back to the Mesolithic era (10,000 to 5,000 B.C.).
By then, *Homo sapiens* had been around for quite a while. I mean, us!
We know that the bones we've found were amputated, because they were cut too cleanly to be caused by an accident. An arm that's been mangled by a *Smilodon*, for example, won't have a clean cut.
In Buthiers-Boulancourt, France, we found evidence of a surgery dating as far back as 6,900 years ago!
What's more: the patient survived! (We found evidence of scarring.)
The man was even fairly old when he died. So, contrary to what you might think, the weak and disabled weren't always abandoned back then.

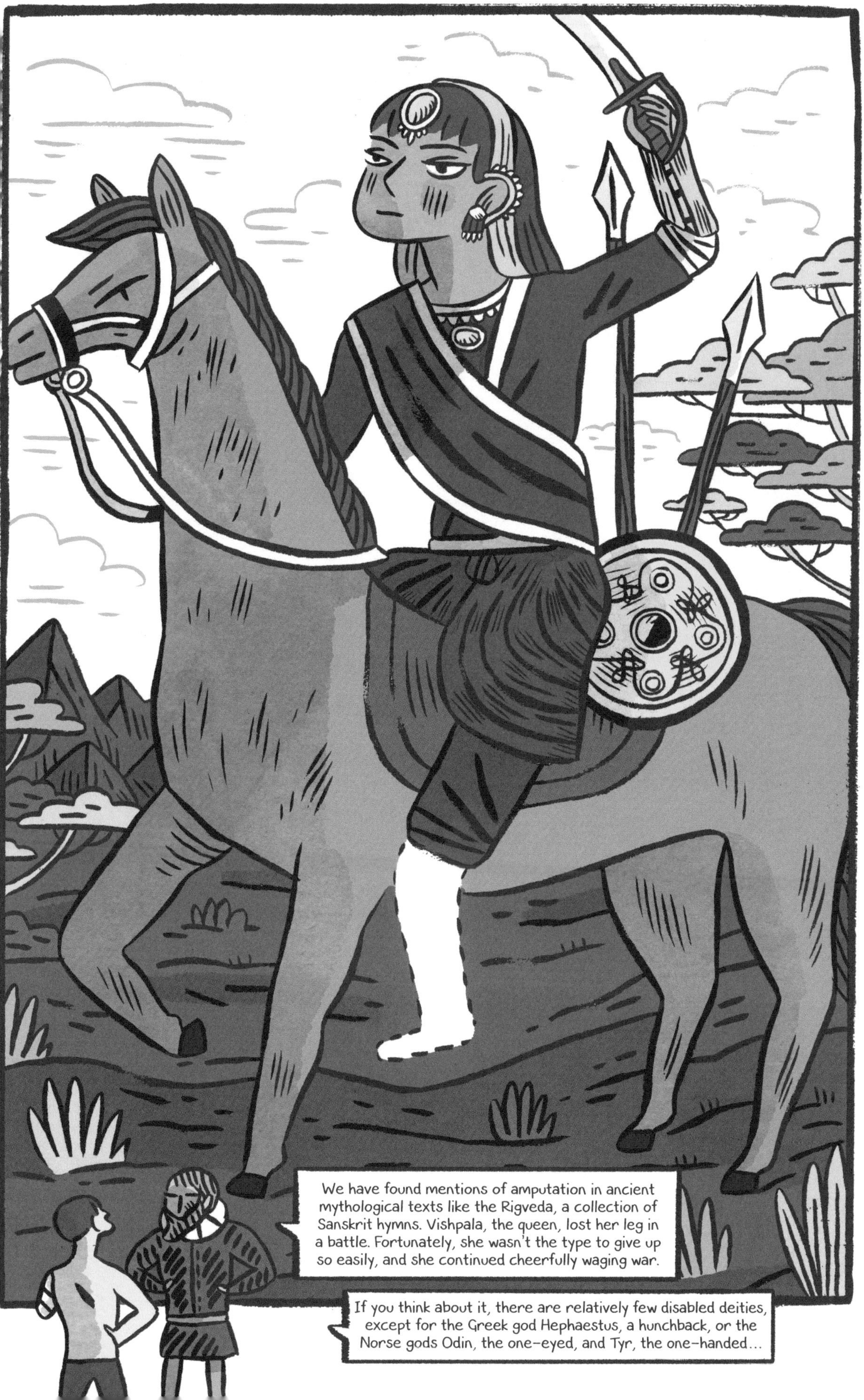

We have found mentions of amputation in ancient mythological texts like the Rigveda, a collection of Sanskrit hymns. Vishpala, the queen, lost her leg in a battle. Fortunately, she wasn't the type to give up so easily, and she continued cheerfully waging war.
If you think about it, there are relatively few disabled deities, except for the Greek god Hephaestus, a hunchback, or the Norse gods Odin, the one-eyed, and Tyr, the one-handed...

THE ANCIENTS

HIPPOCRATES

CELSUS

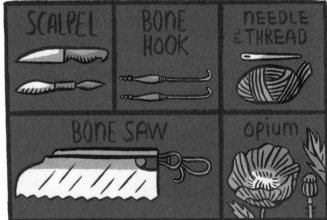

SCALPEL

BONE HOOK

NEEDLE & THREAD

BONE SAW

opium

This period produced two great figures in the history of medicine.

The Greek Hippocrates, who cut out gangrene and tied up blood vessels to stop hemorrhages.

And the Roman Aulus Cornelius Celsus, the author of texts that remain our best sources of information on ancient medicine.

Amputation in the Medieval Era
In the Middle Ages, it was quick and dirty: You cut...
AAAAAAAAAA
AAAAAAAAA
you dunk the stump in boiling oil...
PSSSHHH
AAAA
AAAA
AAAAAAA
AAAAA
...and you seal with hot iron.
HOU HOU
HOU HOU HOU
HOU HOU HOU
It's great for population control.
THE RESULTS
INFECTION
BLOOD LOSS
DEATH

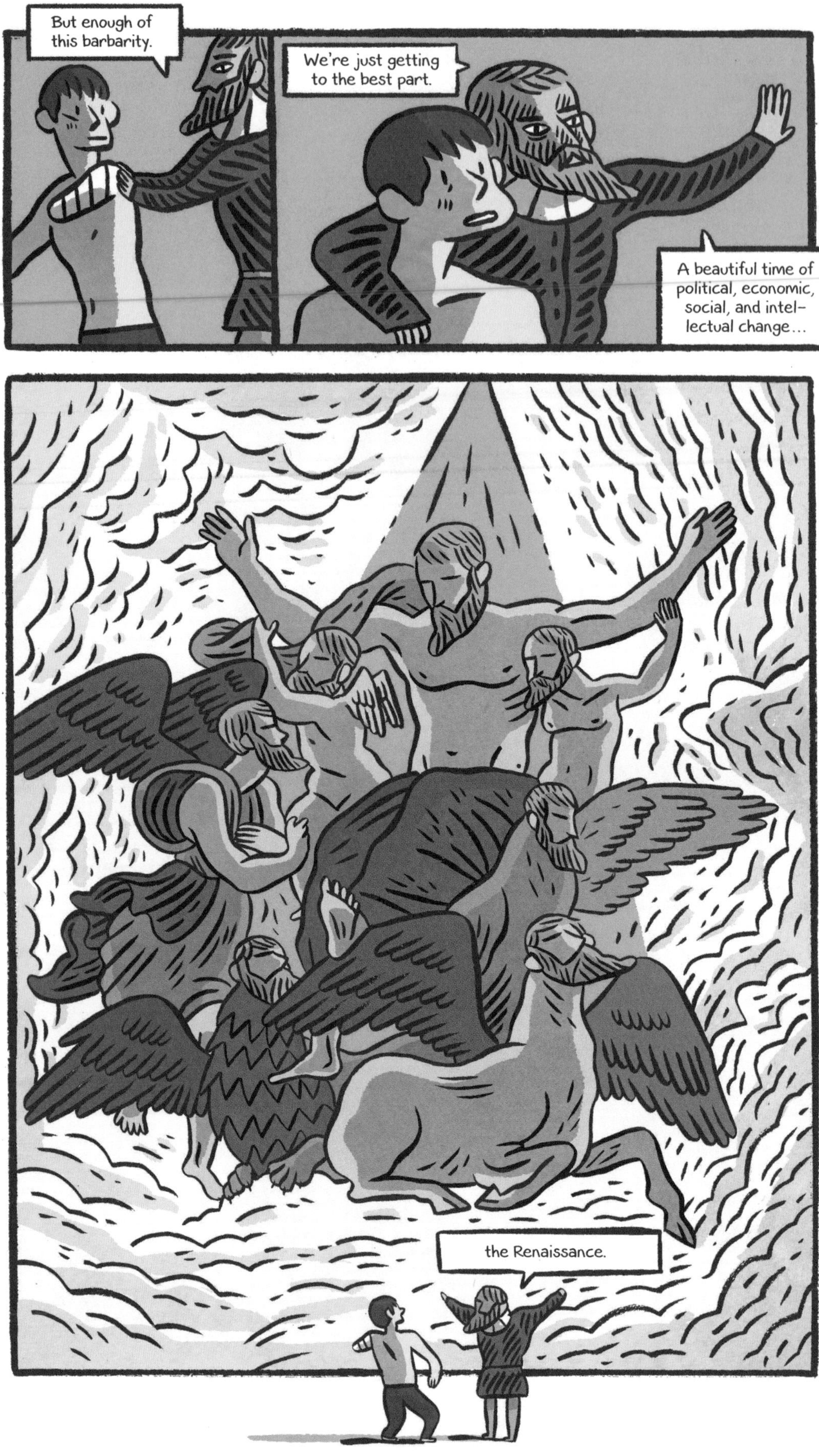
But enough of this barbarity.
We're just getting to the best part.
A beautiful time of political, economic, social, and intel-lectual change...
the Renaissance.

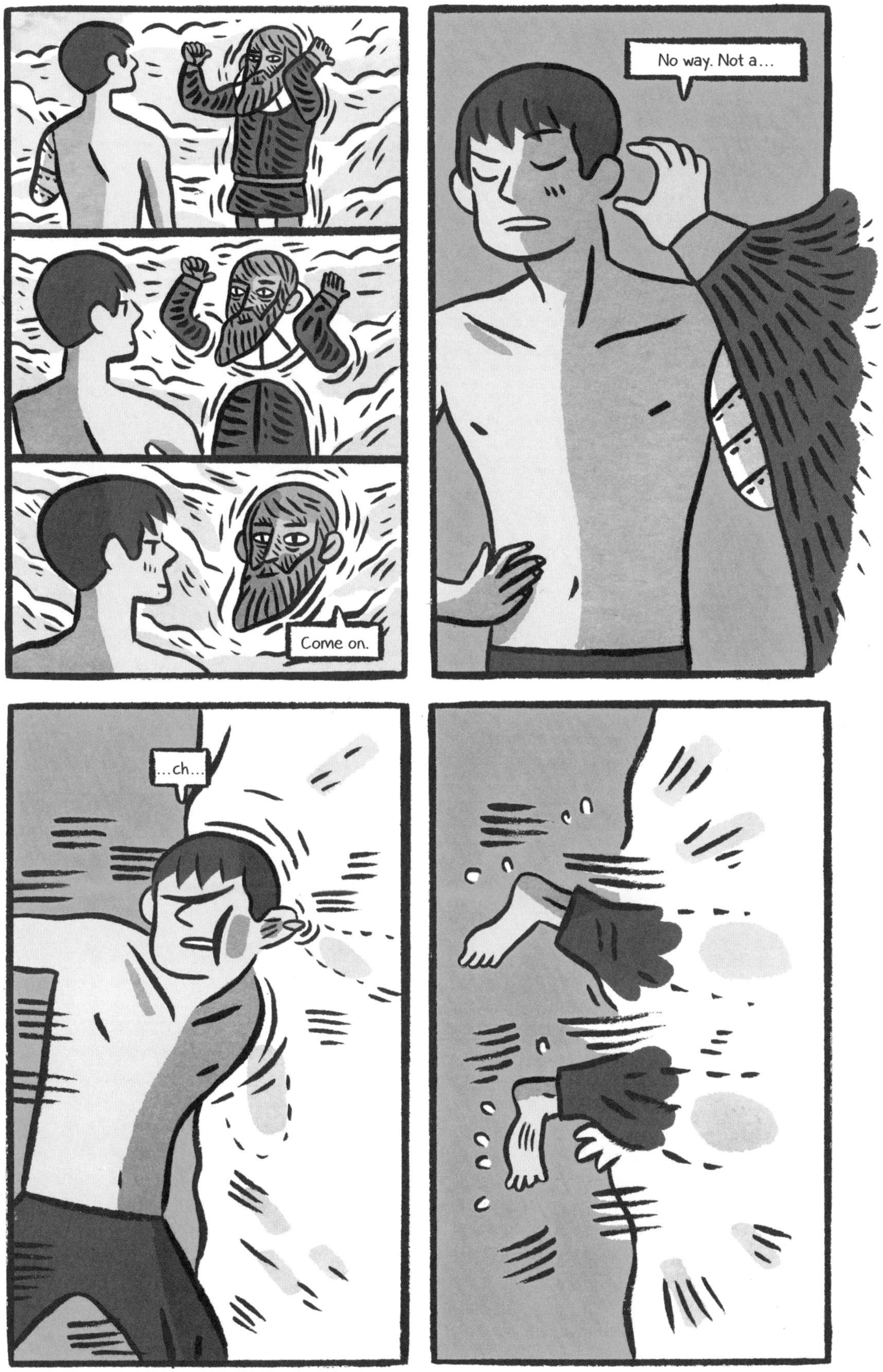
Come on.
No way. Not a…
…ch…

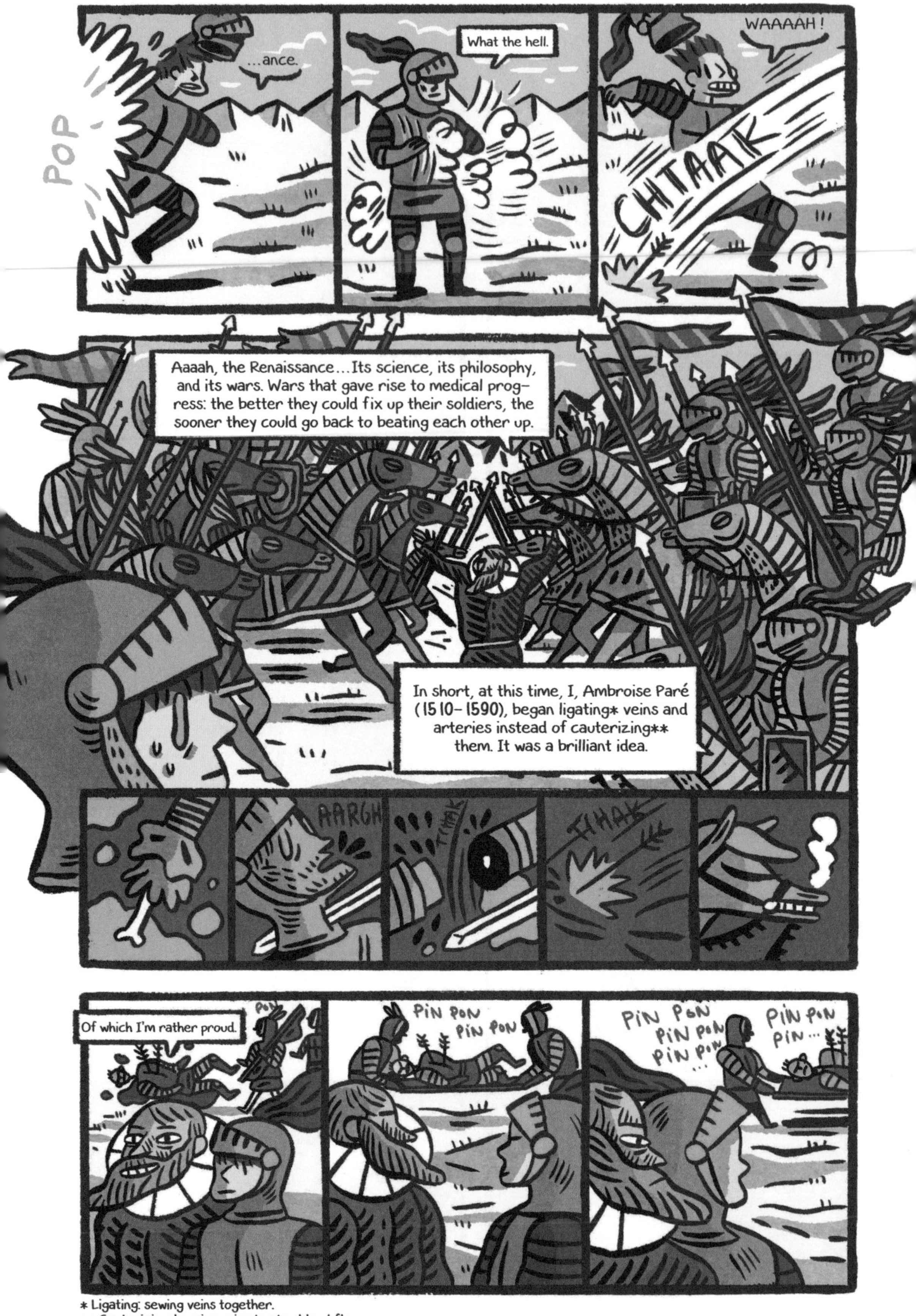

POP
...ance.
What the hell.
WAAAAH!
CHTAAK
Aaaah, the Renaissance... Its science, its philosophy, and its wars. Wars that gave rise to medical progress: the better they could fix up their soldiers, the sooner they could go back to beating each other up.
In short, at this time, I, Ambroise Paré (1510–1590), began ligating* veins and arteries instead of cauterizing** them. It was a brilliant idea.
AARGH
Of which I'm rather proud.
PIN PON PIN PON
PIN PON PIN PON PIN PON...
PIN PON PIN...

* Ligating: sewing veins together.
** Cauterizing: burning veins to stop blood flow.

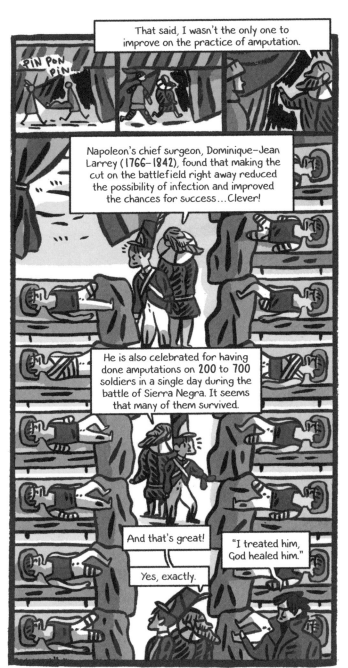

That said, I wasn't the only one to improve on the practice of amputation.

PIN PON PIN...

Napoleon's chief surgeon, Dominique-Jean Larrey (1766–1842), found that making the cut on the battlefield right away reduced the possibility of infection and improved the chances for success...Clever!

He is also celebrated for having done amputations on 200 to 700 soldiers in a single day during the battle of Sierra Negra. It seems that many of them survived.

And that's great!

Yes, exactly.

"I treated him, God healed him."

Is the costume really necessary?

Yep, so you'll blend in and be more emotionally invested.

Awesome.

Come on.

You need to see this.

The work of another underappreciated genius of this era is crucial for the history of medicine.

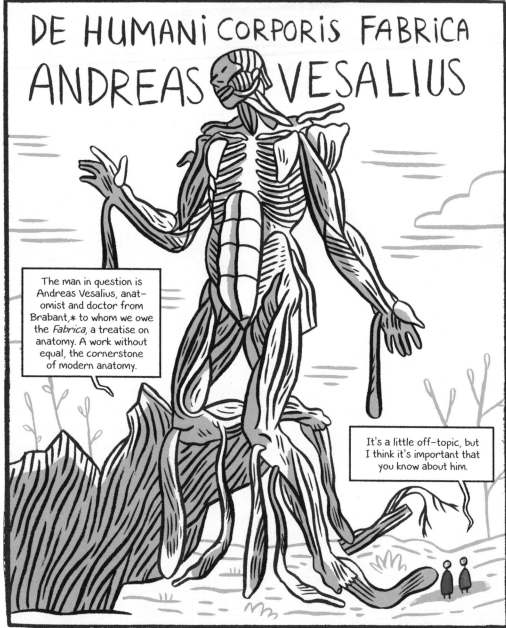

DE HUMANI CORPORIS FABRICA
ANDREAS VESALIUS

The man in question is Andreas Vesalius, anatomist and doctor from Brabant,* to whom we owe the *Fabrica*, a treatise on anatomy. A work without equal, the cornerstone of modern anatomy.

It's a little off-topic, but I think it's important that you know about him.

* A duchy in Belgium.

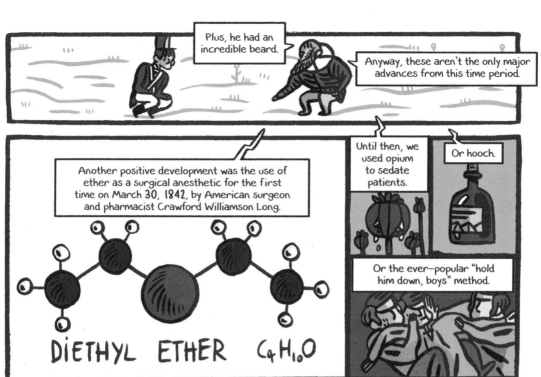

Plus, he had an incredible beard.

Anyway, these aren't the only major advances from this time period.

Another positive development was the use of ether as a surgical anesthetic for the first time on March 30, 1842, by American surgeon and pharmacist Crawford Williamson Long.

DIETHYL ETHER $C_4H_{10}O$

Until then, we used opium to sedate patients.

Or hooch.

Or the ever-popular "hold him down, boys" method.

Of course, we can't forget Louis Pasteur, whose work revolutionized antisepsis.

Since the 17th century, we had been conscious of the microbes and bacteria present all around us and of their role in infections.

We'd gone from the concept of antisepsis (fighting infections) to that of asepsis (preventing infections).

It was only after Pasteur that we began to boil instruments and avoid putting our big, dirty fingers in wounds.

Strangely, there have been very few infections since…

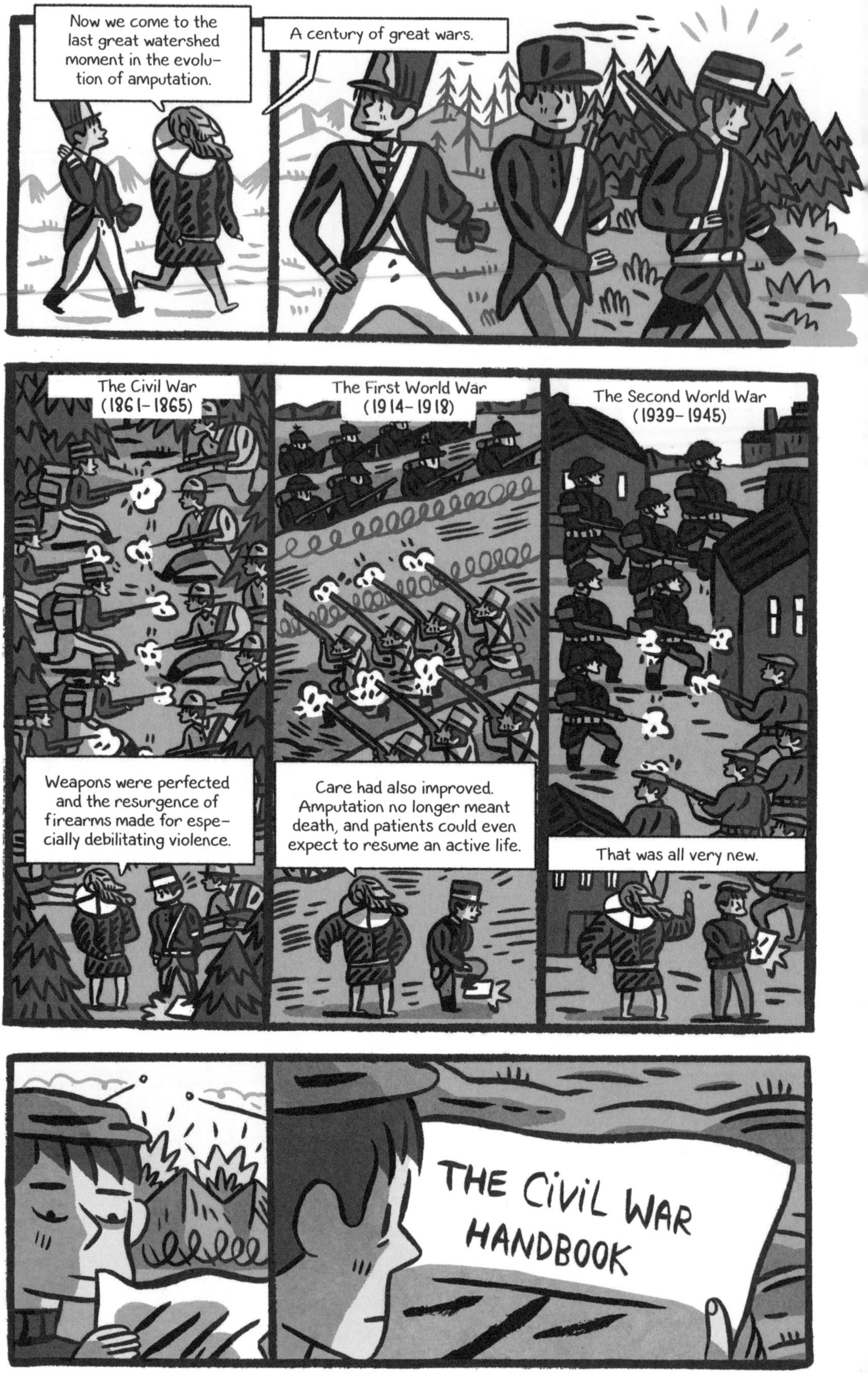
Now we come to the last great watershed moment in the evolution of amputation.
A century of great wars.
The Civil War (1861–1865)
The First World War (1914–1918)
The Second World War (1939–1945)
Weapons were perfected and the resurgence of firearms made for especially debilitating violence.
Care had also improved. Amputation no longer meant death, and patients could even expect to resume an active life.
That was all very new.
THE CIVIL WAR HANDBOOK

You too can properly amputate, thanks to the

TOURNIQUET !

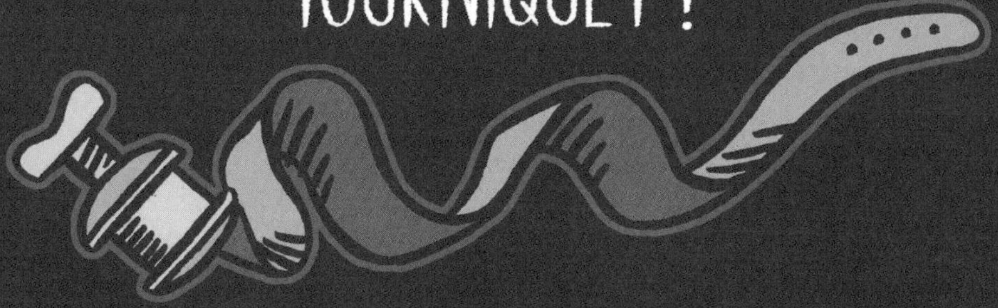

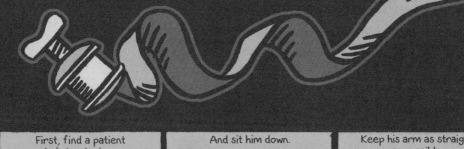

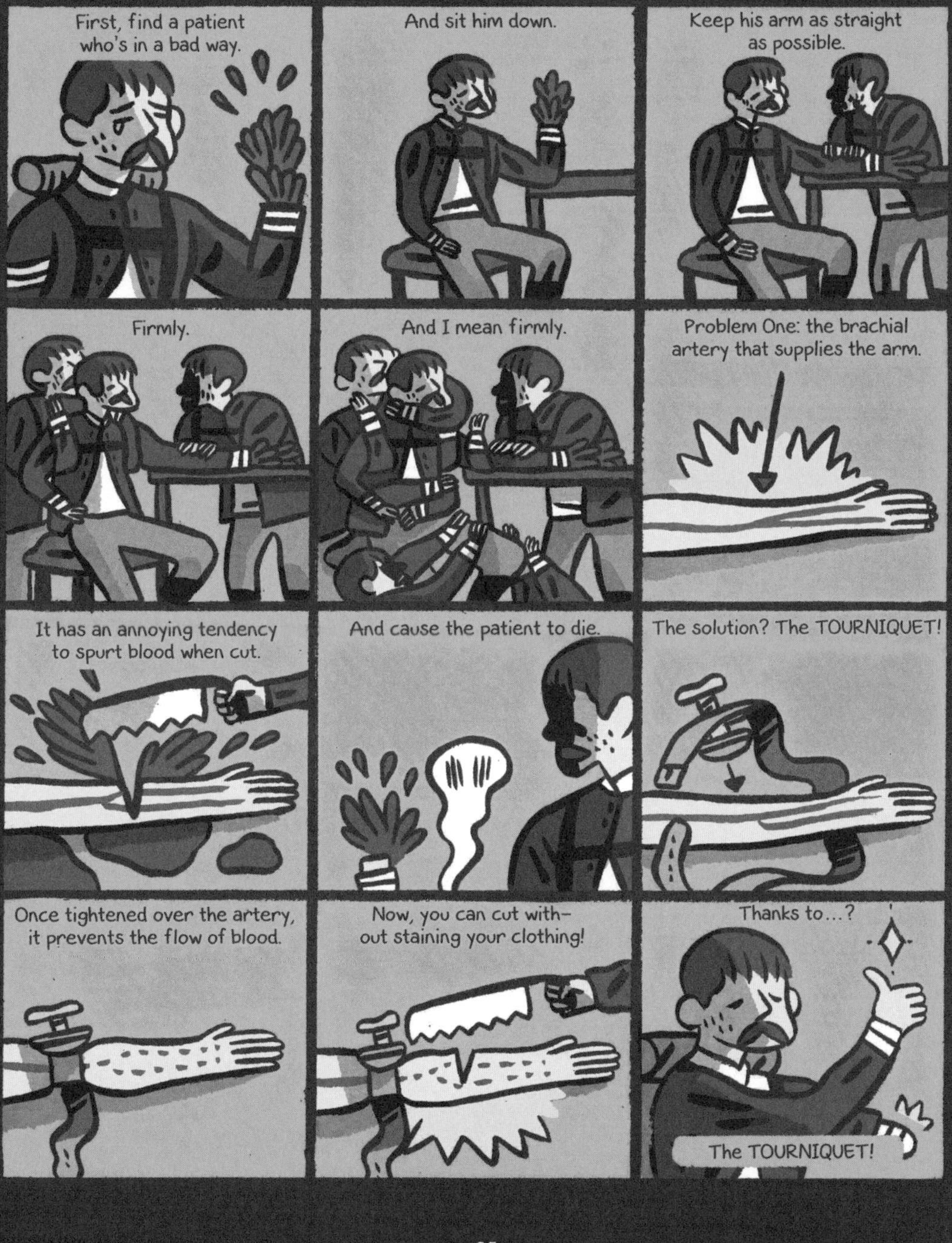
First, find a patient who's in a bad way.
And sit him down.
Keep his arm as straight as possible.
Firmly.
And I mean firmly.
Problem One: the brachial artery that supplies the arm.
It has an annoying tendency to spurt blood when cut.
And cause the patient to die.
The solution? The TOURNIQUET!
Once tightened over the artery, it prevents the flow of blood.
Now, you can cut with-out staining your clothing!
Thanks to...?
The TOURNIQUET!

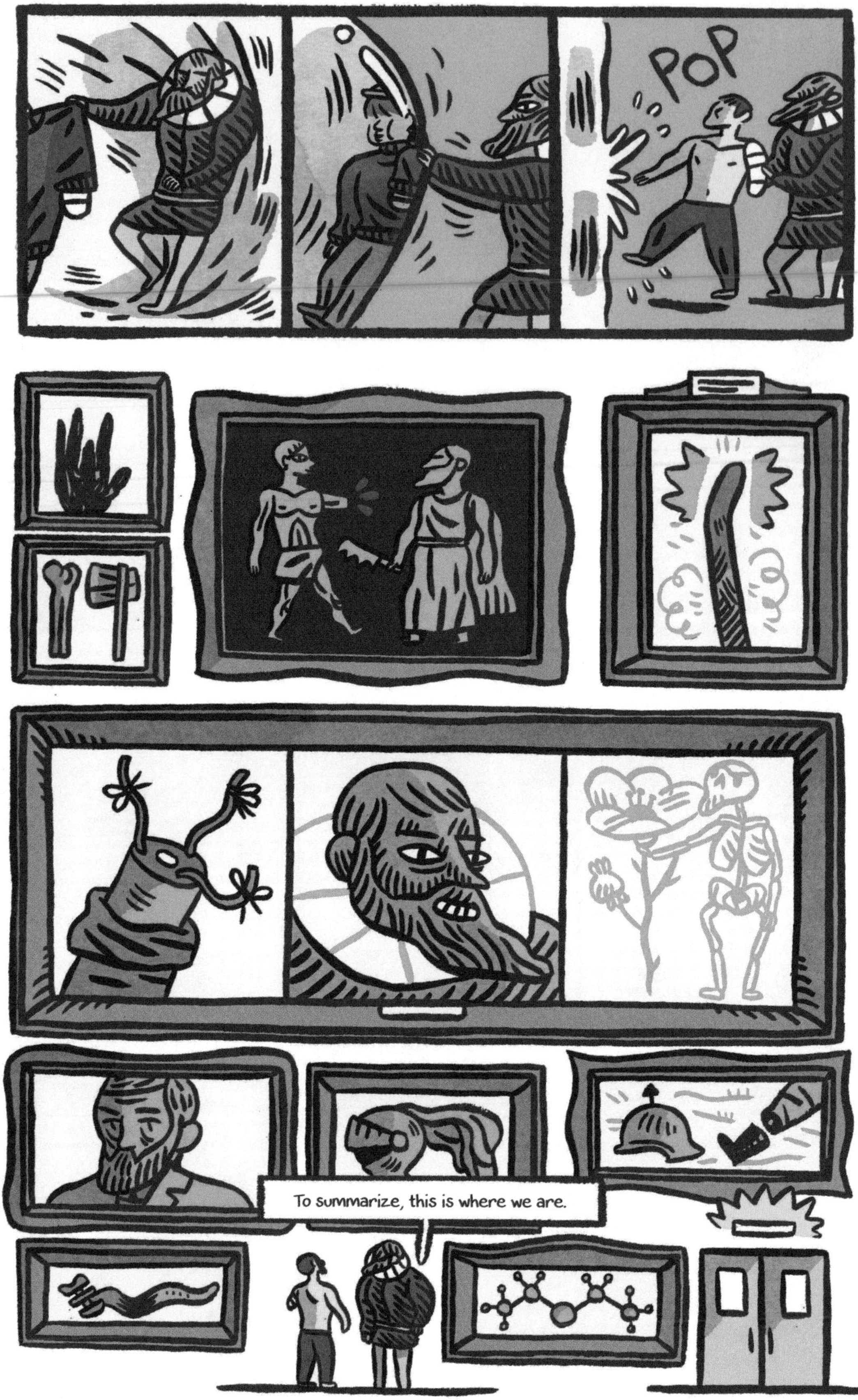
POP
To summarize, this is where we are.

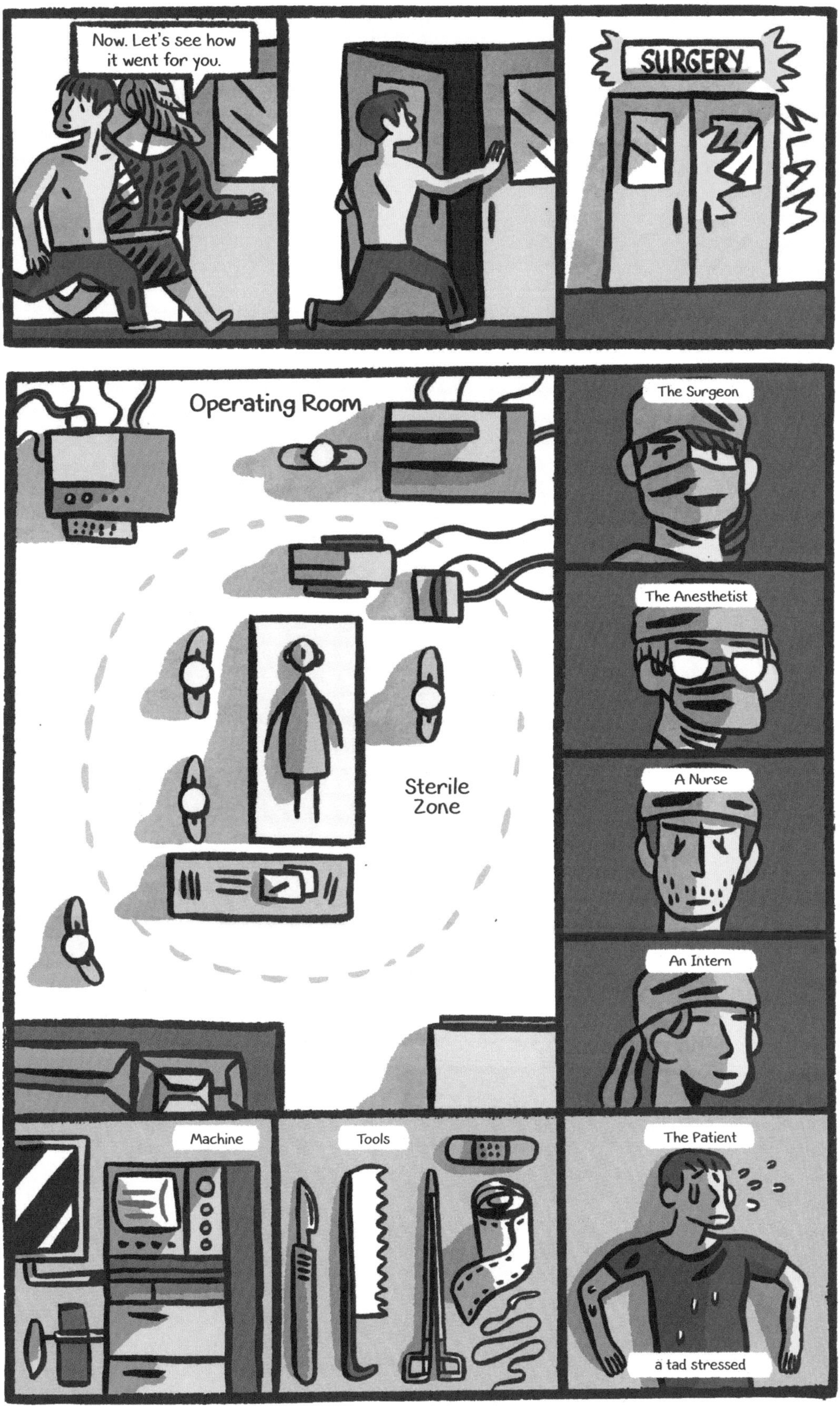
Now. Let's see how it went for you.
SURGERY
SLAM
Operating Room
Sterile Zone
The Surgeon
The Anesthetist
A Nurse
An Intern
Machine
Tools
The Patient
a tad stressed

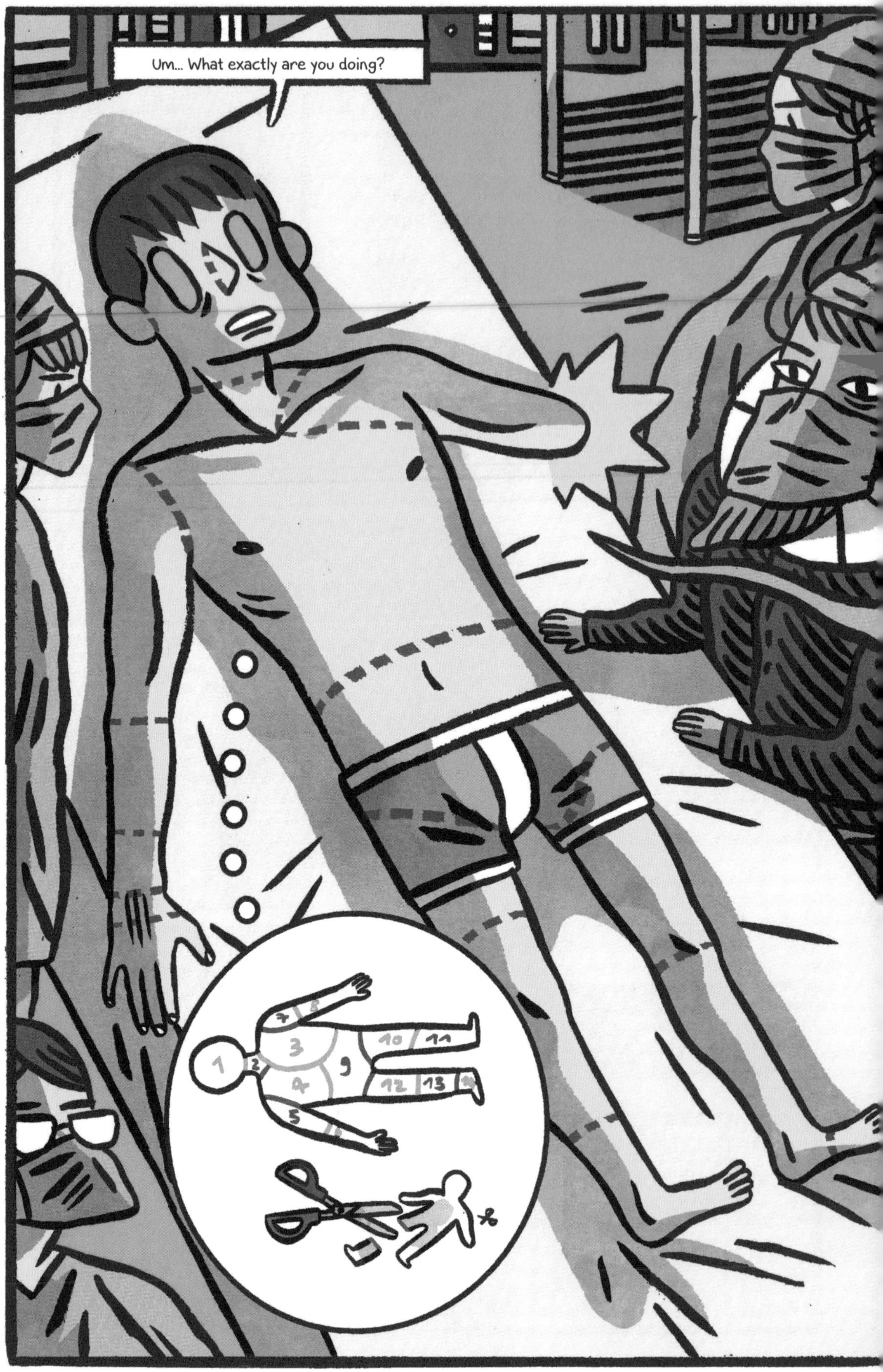
Um... What exactly are you doing?

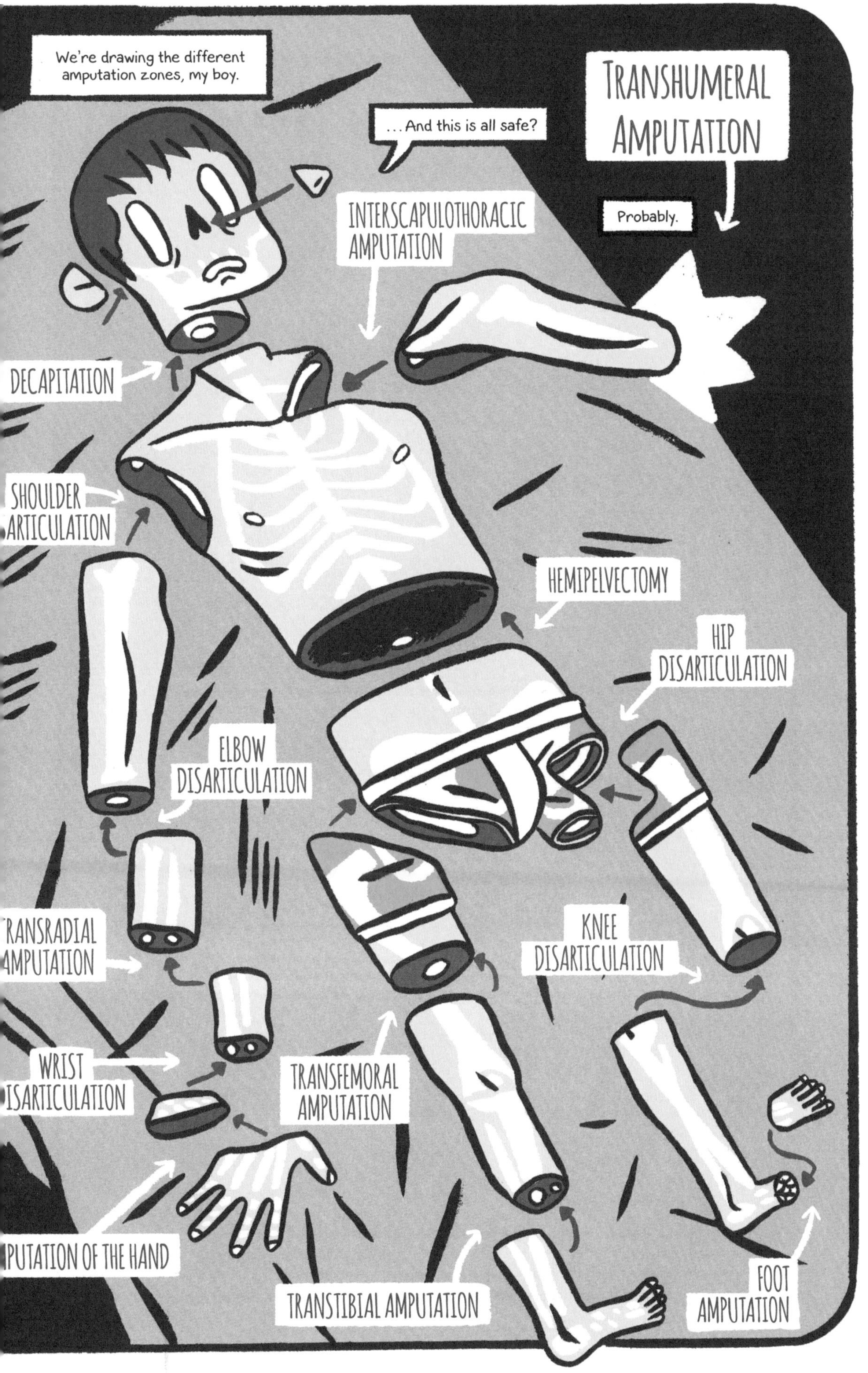

We're drawing the different amputation zones, my boy.
. . . And this is all safe?
TRANSHUMERAL AMPUTATION
Probably.
INTERSCAPULOTHORACIC AMPUTATION
DECAPITATION
SHOULDER ARTICULATION
HEMIPELVECTOMY
HIP DISARTICULATION
ELBOW DISARTICULATION
RANSRADIAL AMPUTATION
KNEE DISARTICULATION
WRIST ISARTICULATION
TRANSFEMORAL AMPUTATION
PUTATION OF THE HAND
TRANSTIBIAL AMPUTATION
FOOT AMPUTATION

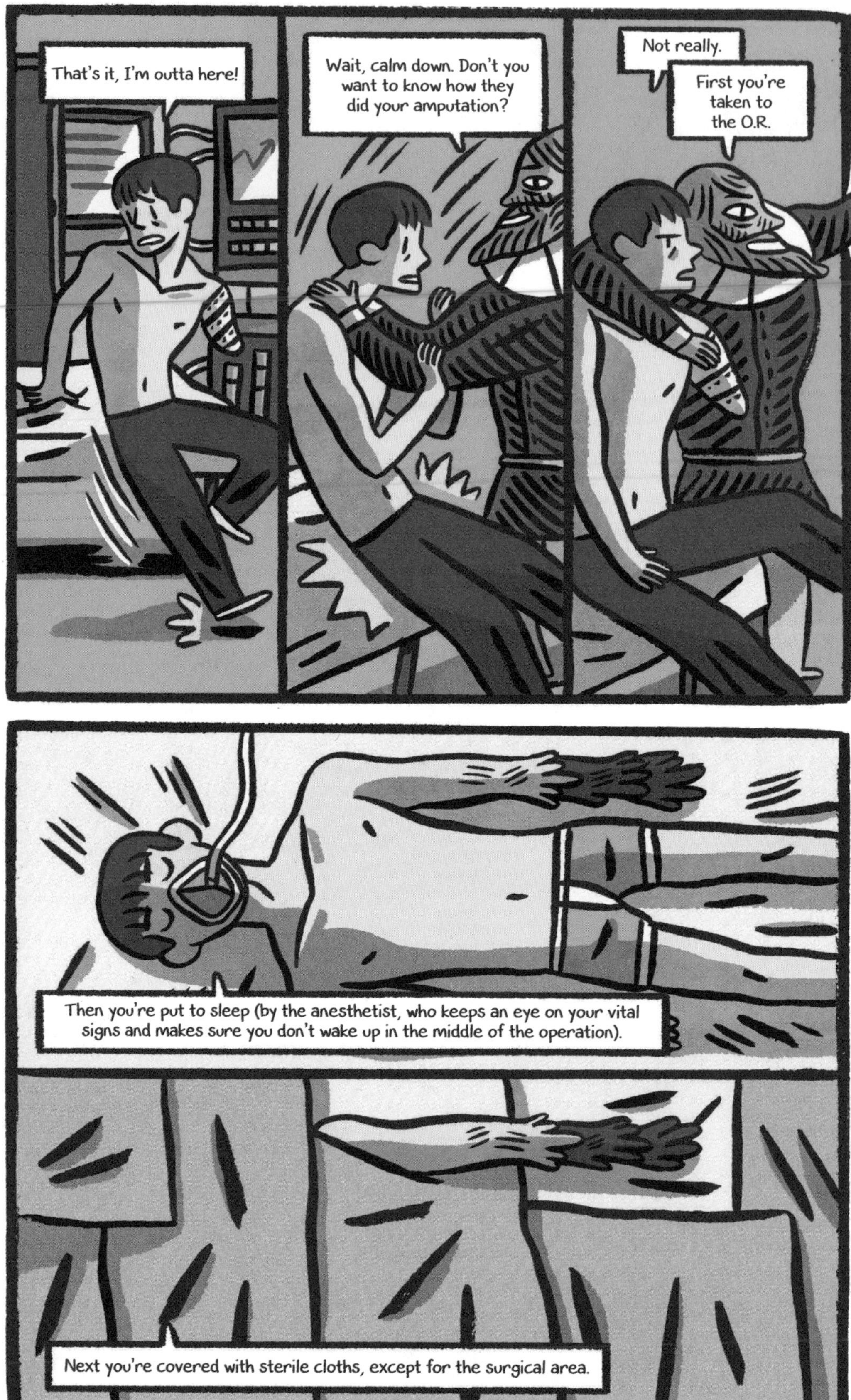

That's it, I'm outta here!
Wait, calm down. Don't you want to know how they did your amputation?
Not really.
First you're taken to the O.R.
Then you're put to sleep (by the anesthetist, who keeps an eye on your vital signs and makes sure you don't wake up in the middle of the operation).
Next you're covered with sterile cloths, except for the surgical area.

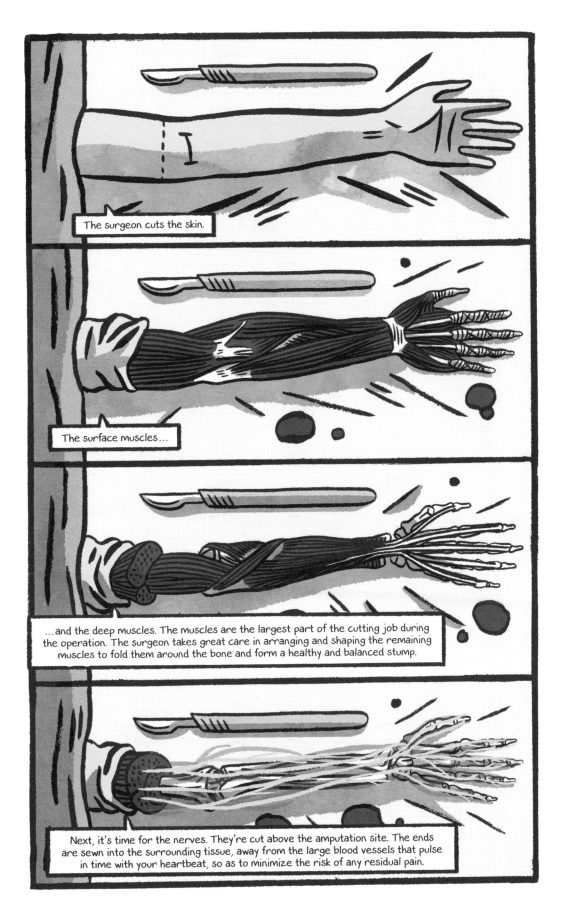

The surgeon cuts the skin.

The surface muscles...

...and the deep muscles. The muscles are the largest part of the cutting job during the operation. The surgeon takes great care in arranging and shaping the remaining muscles to fold them around the bone and form a healthy and balanced stump.

Next, it's time for the nerves. They're cut above the amputation site. The ends are sewn into the surrounding tissue, away from the large blood vessels that pulse in time with your heartbeat, so as to minimize the risk of any residual pain.

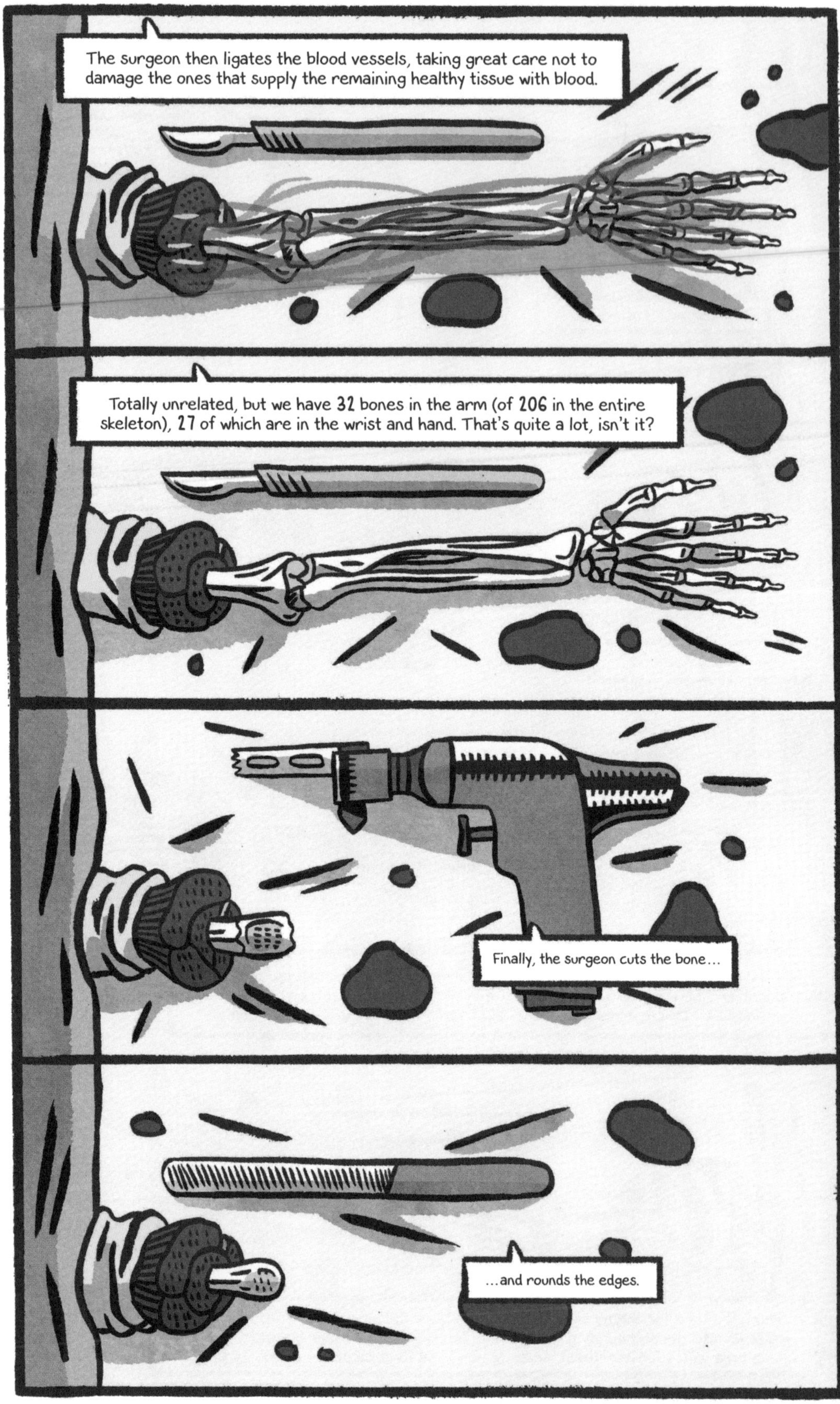
The surgeon then ligates the blood vessels, taking great care not to damage the ones that supply the remaining healthy tissue with blood.
Totally unrelated, but we have 32 bones in the arm (of 206 in the entire skeleton), 27 of which are in the wrist and hand. That's quite a lot, isn't it?
Finally, the surgeon cuts the bone…
…and rounds the edges.

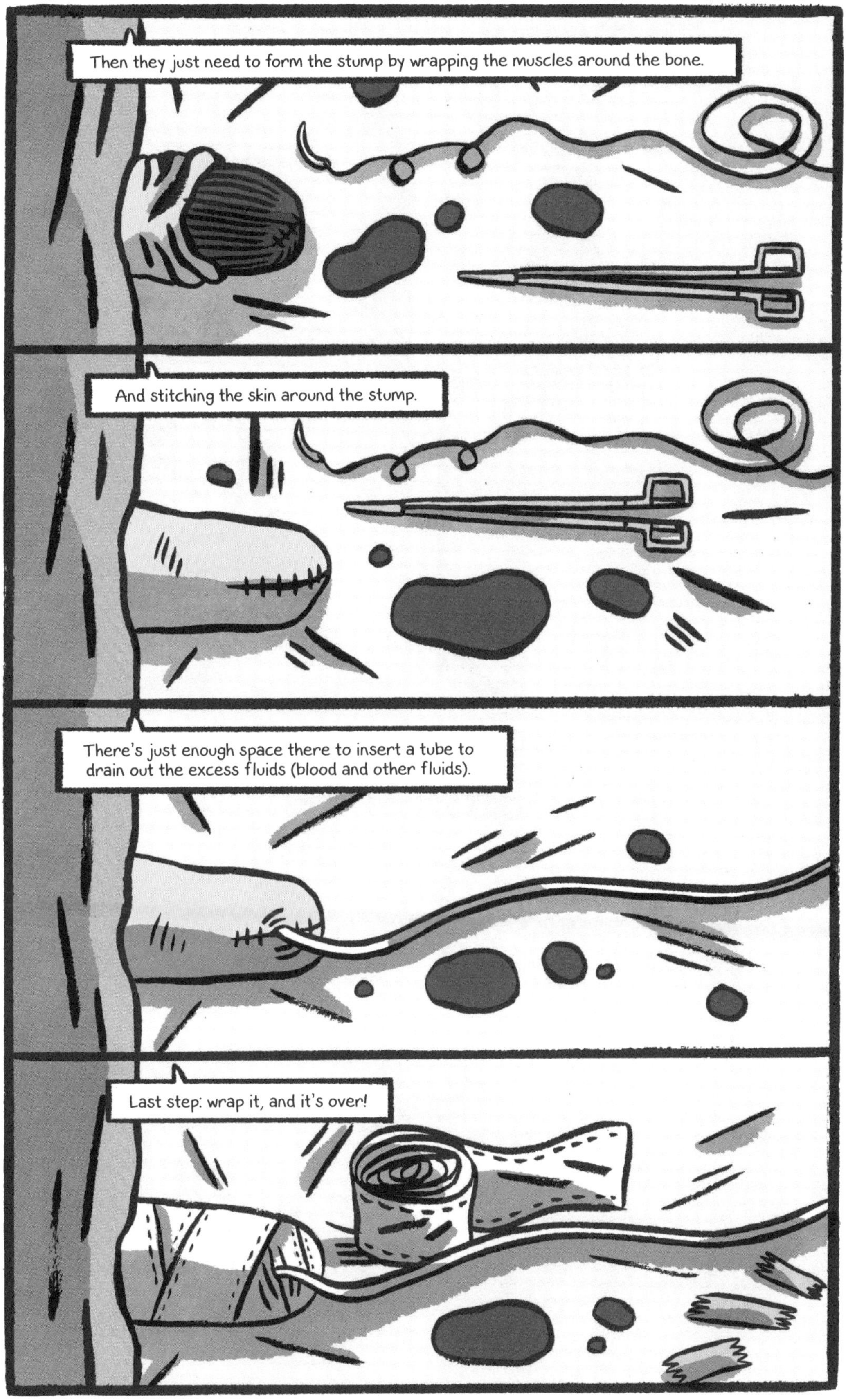
Then they just need to form the stump by wrapping the muscles around the bone.
And stitching the skin around the stump.
There's just enough space there to insert a tube to drain out the excess fluids (blood and other fluids).
Last step: wrap it, and it's over!

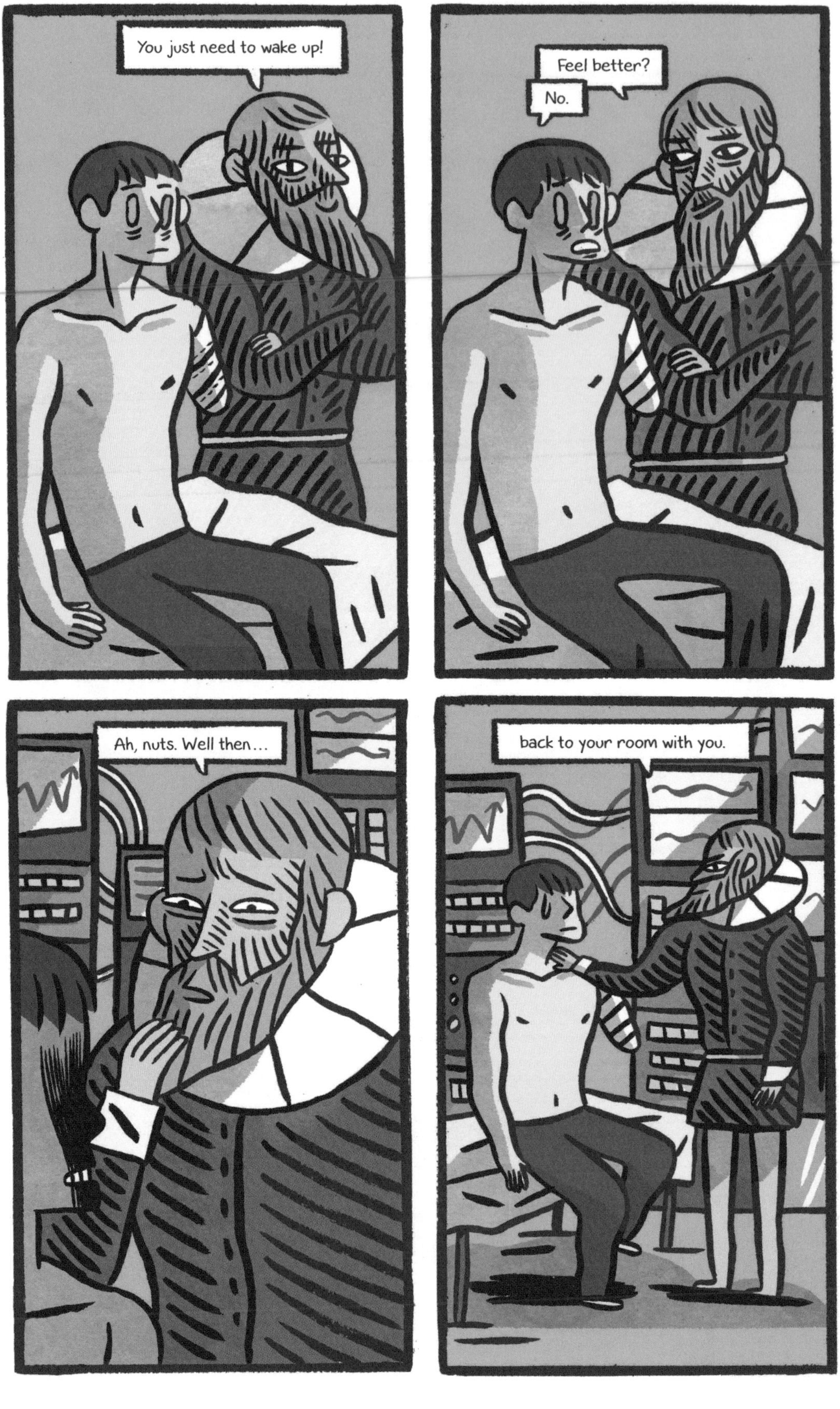
You just need to wake up!
Feel better?
No.
Ah, nuts. Well then...
back to your room with you.

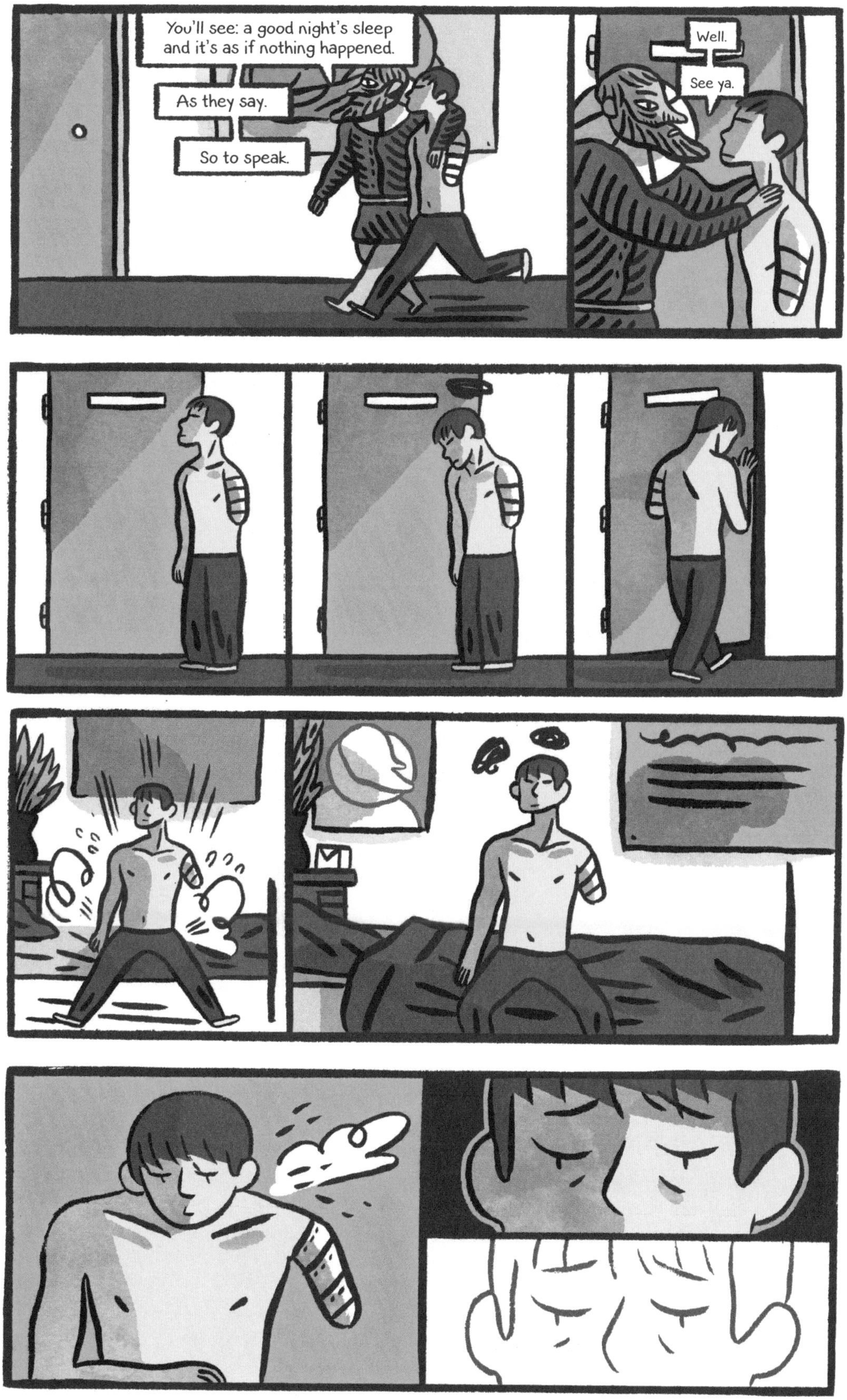
You'll see: a good night's sleep and it's as if nothing happened.
As they say.
So to speak.
Well.
See ya.

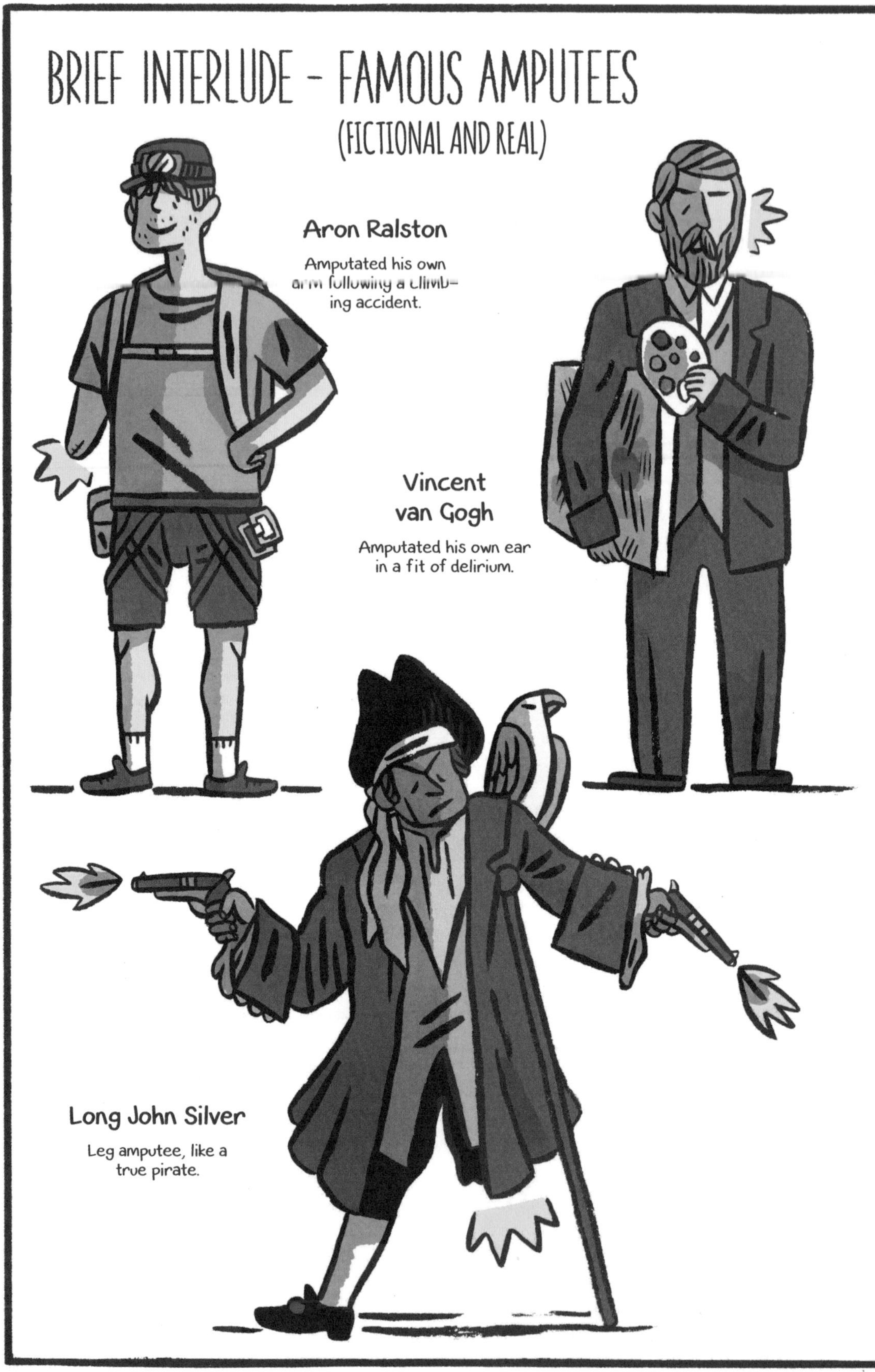
BRIEF INTERLUDE - FAMOUS AMPUTEES
(FICTIONAL AND REAL)
Aron Ralston
Amputated his own arm following a climbing accident.
Vincent van Gogh
Amputated his own ear in a fit of delirium.
Long John Silver
Leg amputee, like a true pirate.

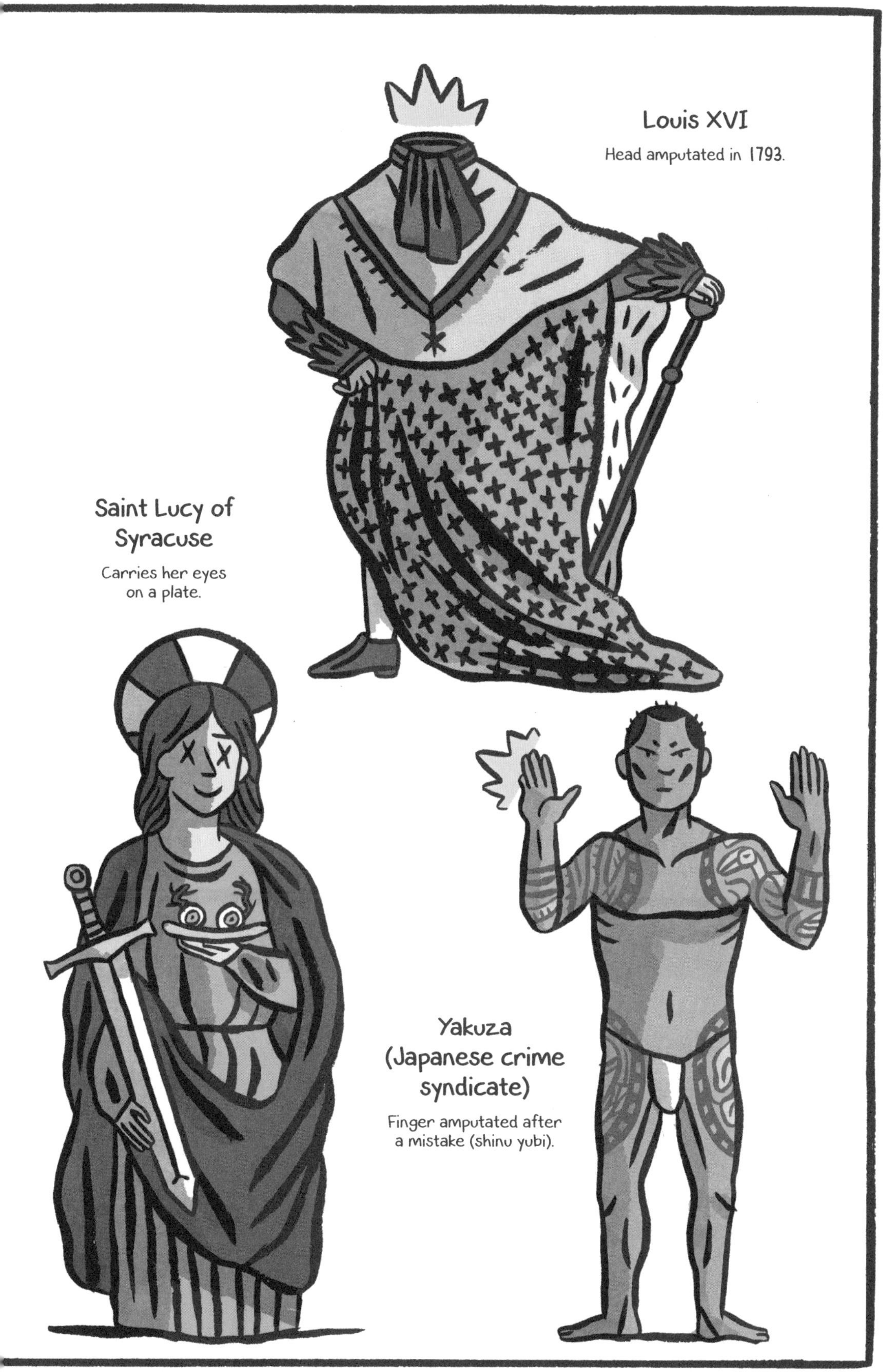
Louis XVI
Head amputated in 1793.
Saint Lucy of Syracuse
Carries her eyes on a plate.
Yakuza (Japanese crime syndicate)
Finger amputated after a mistake (shinu yubi).

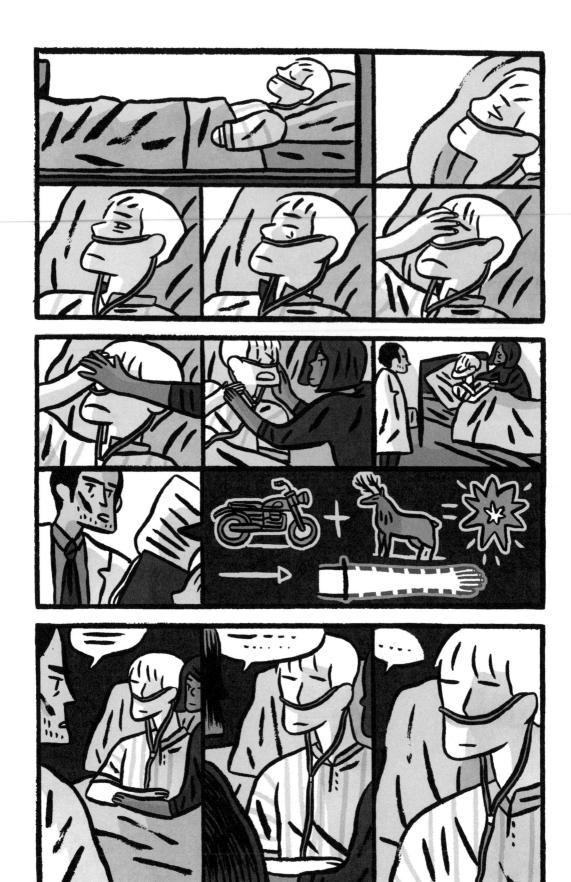

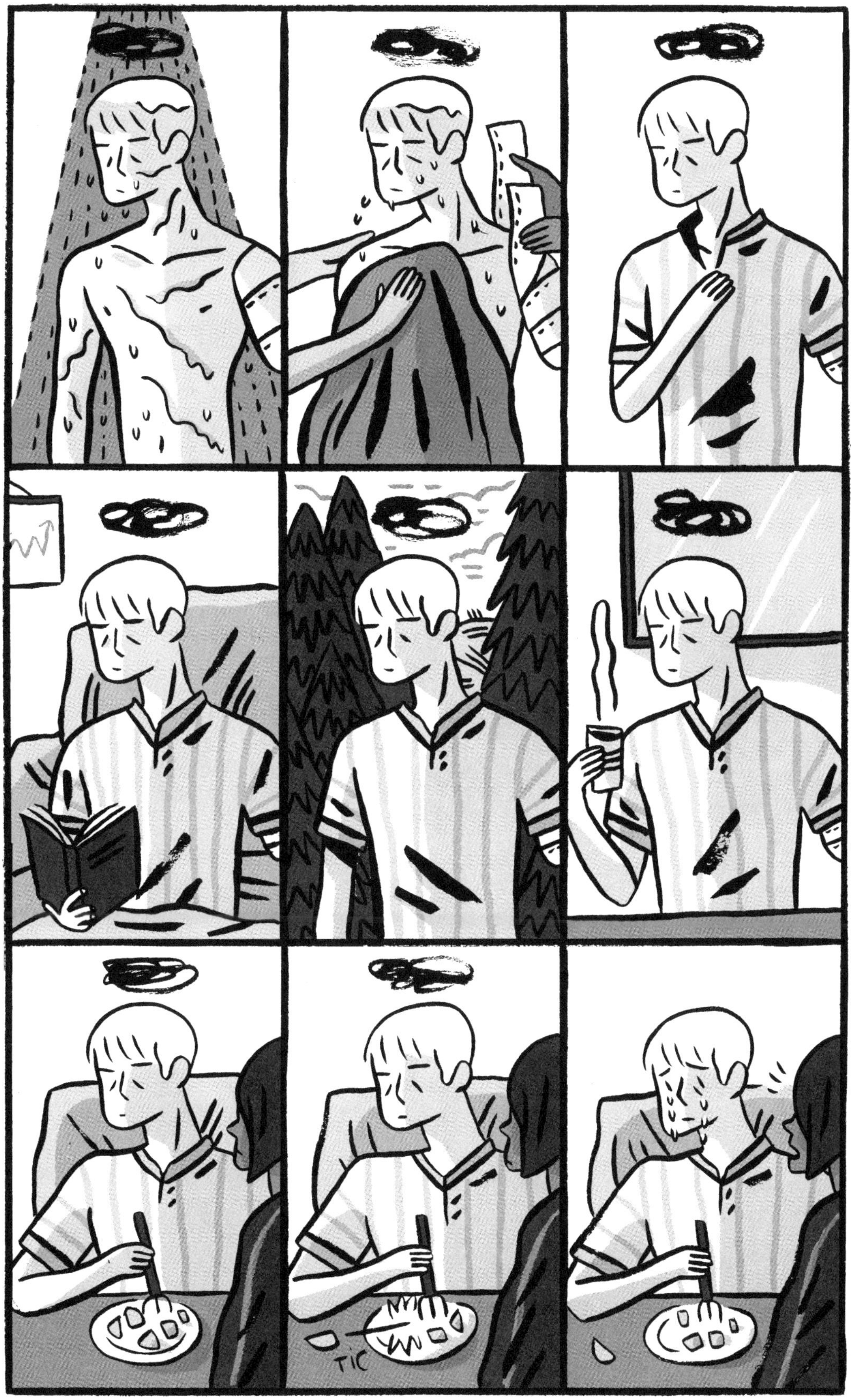
TIC

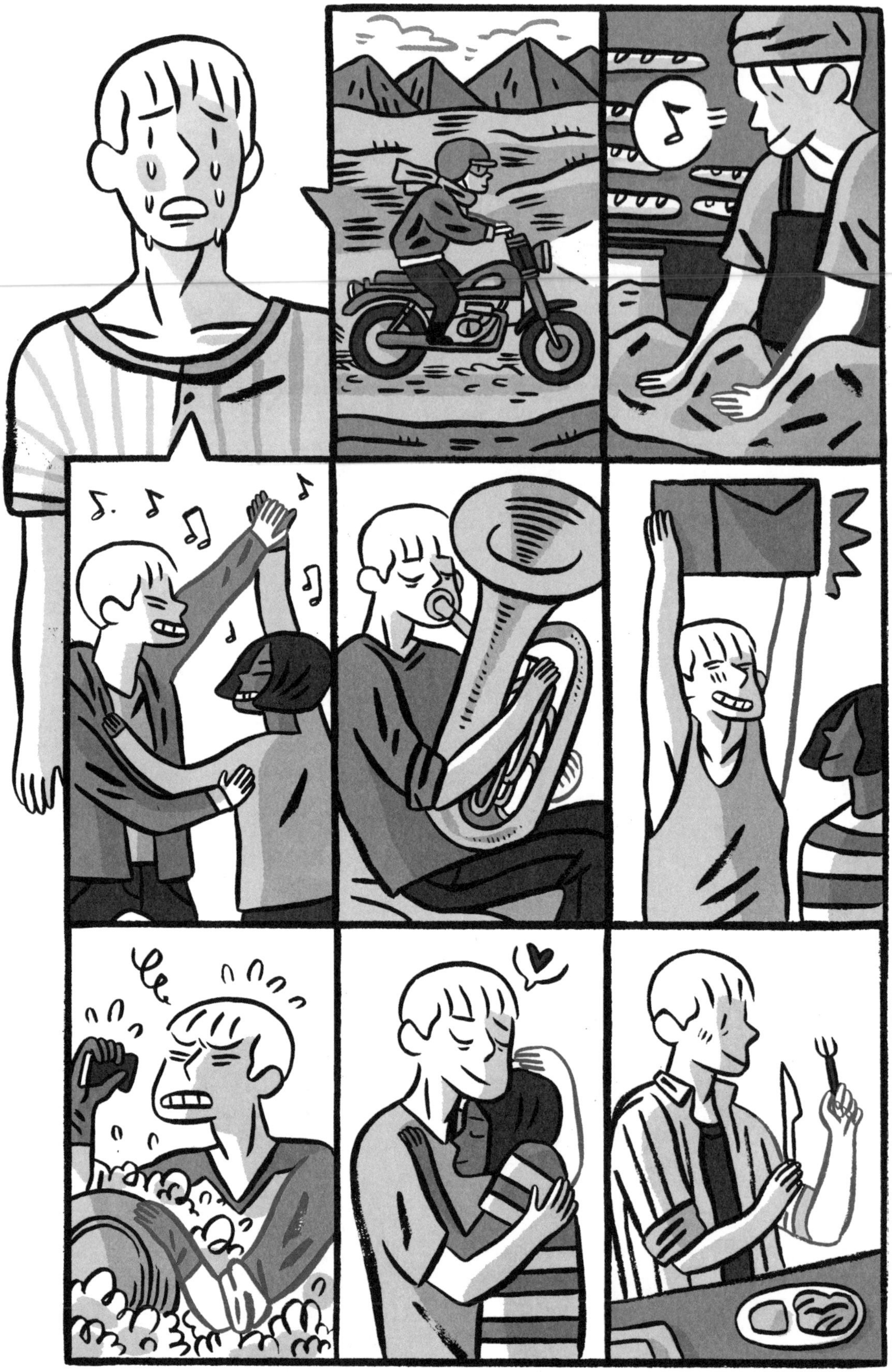

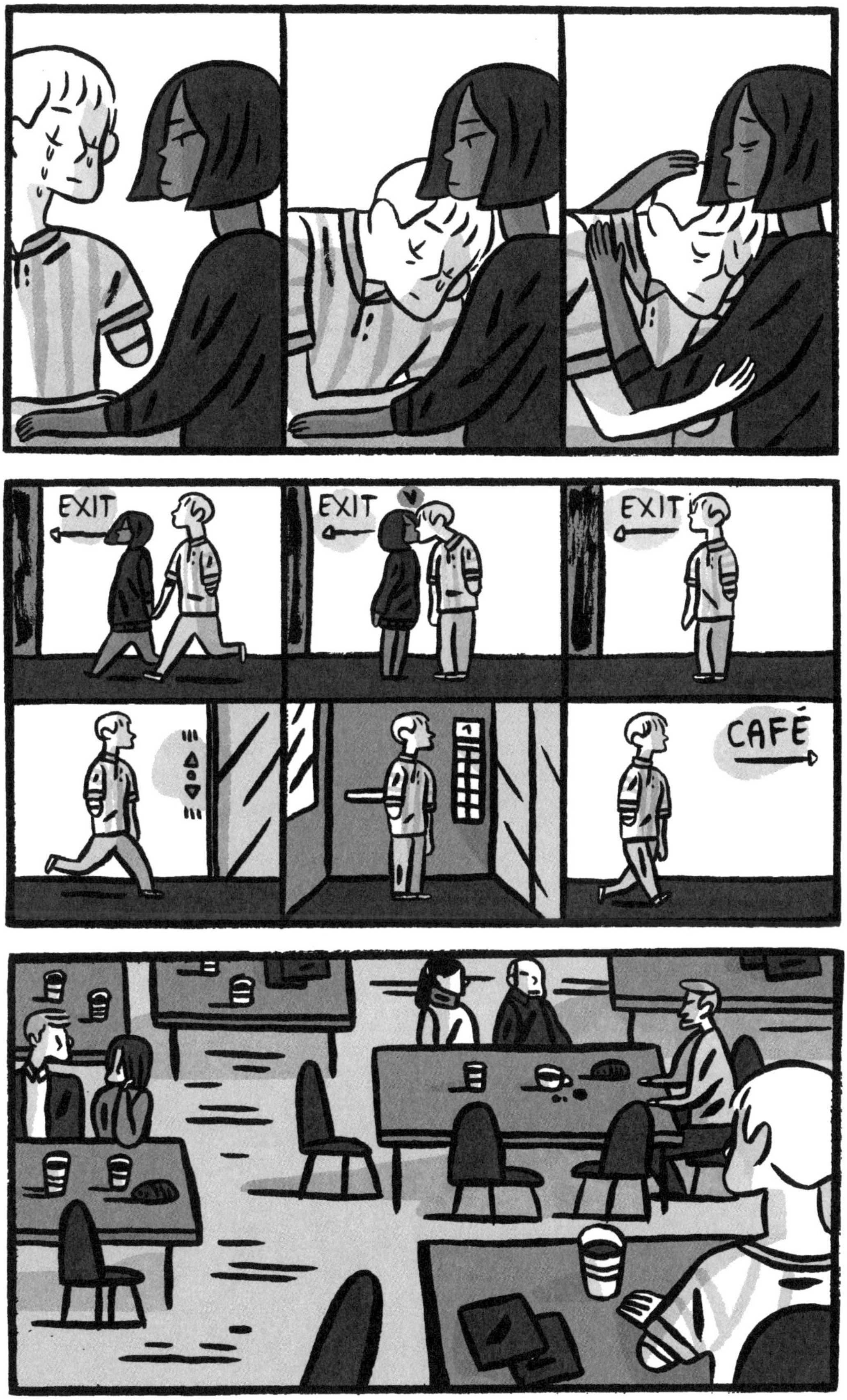
EXIT
EXIT
EXIT
CAFÉ

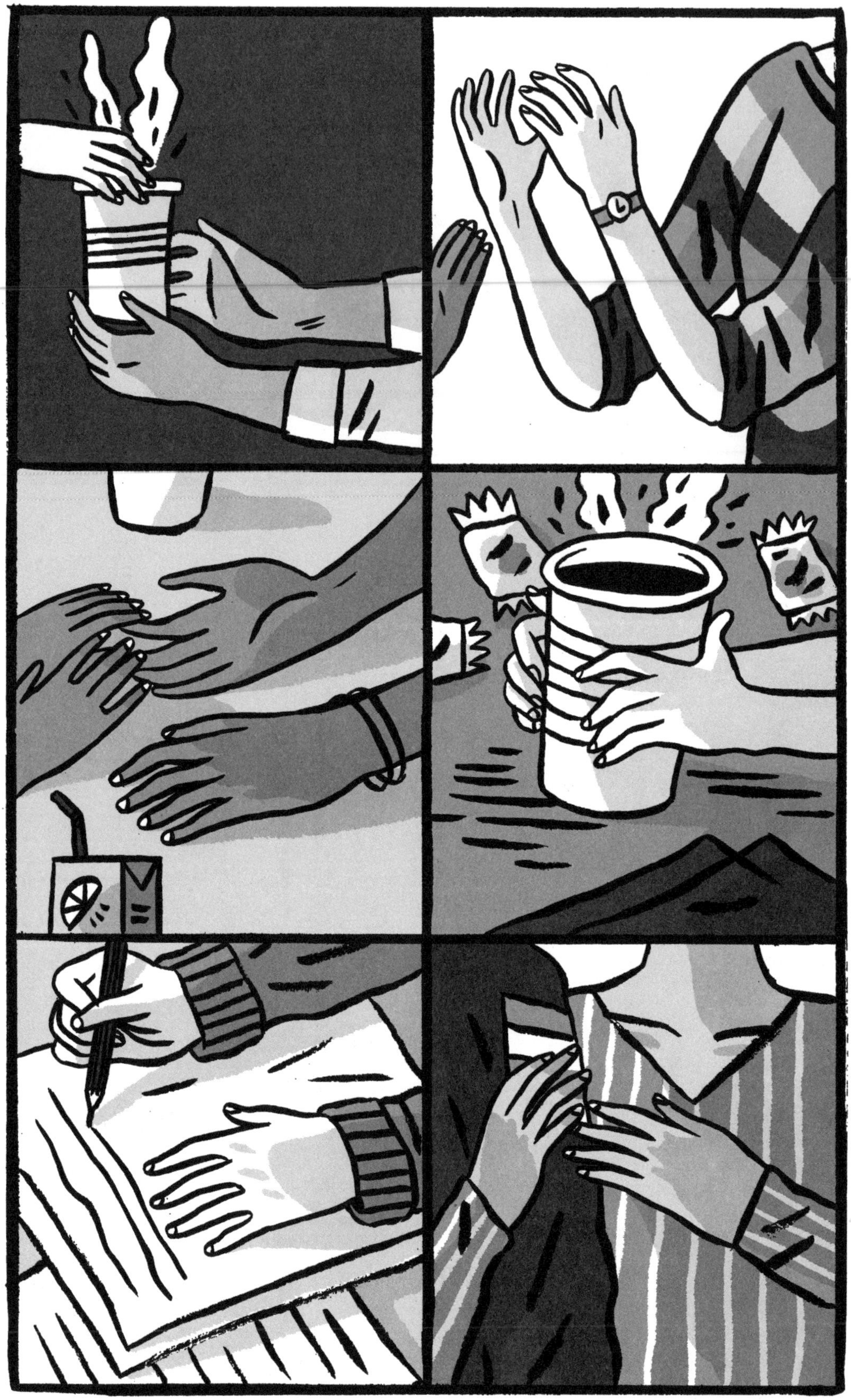

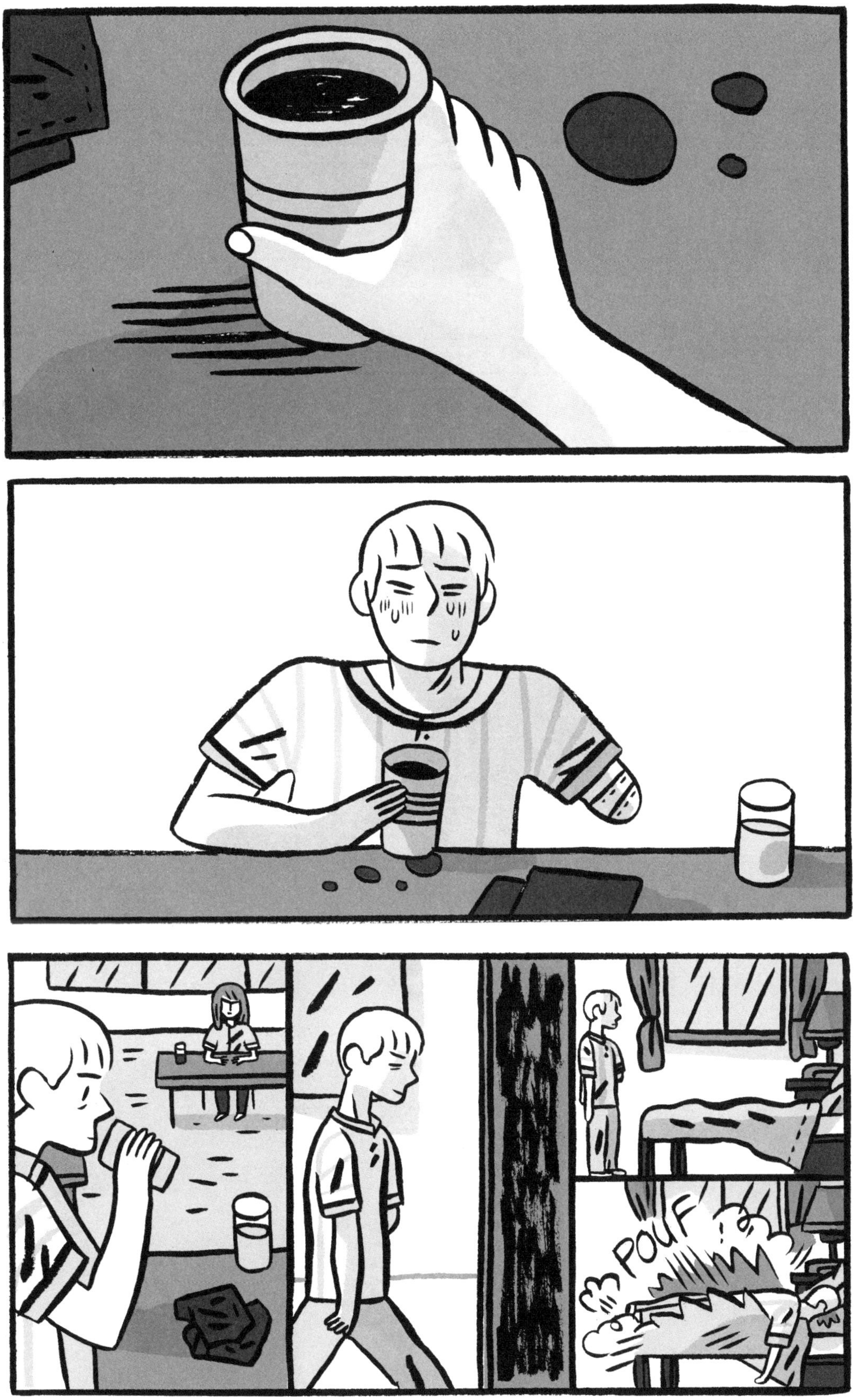
POUF

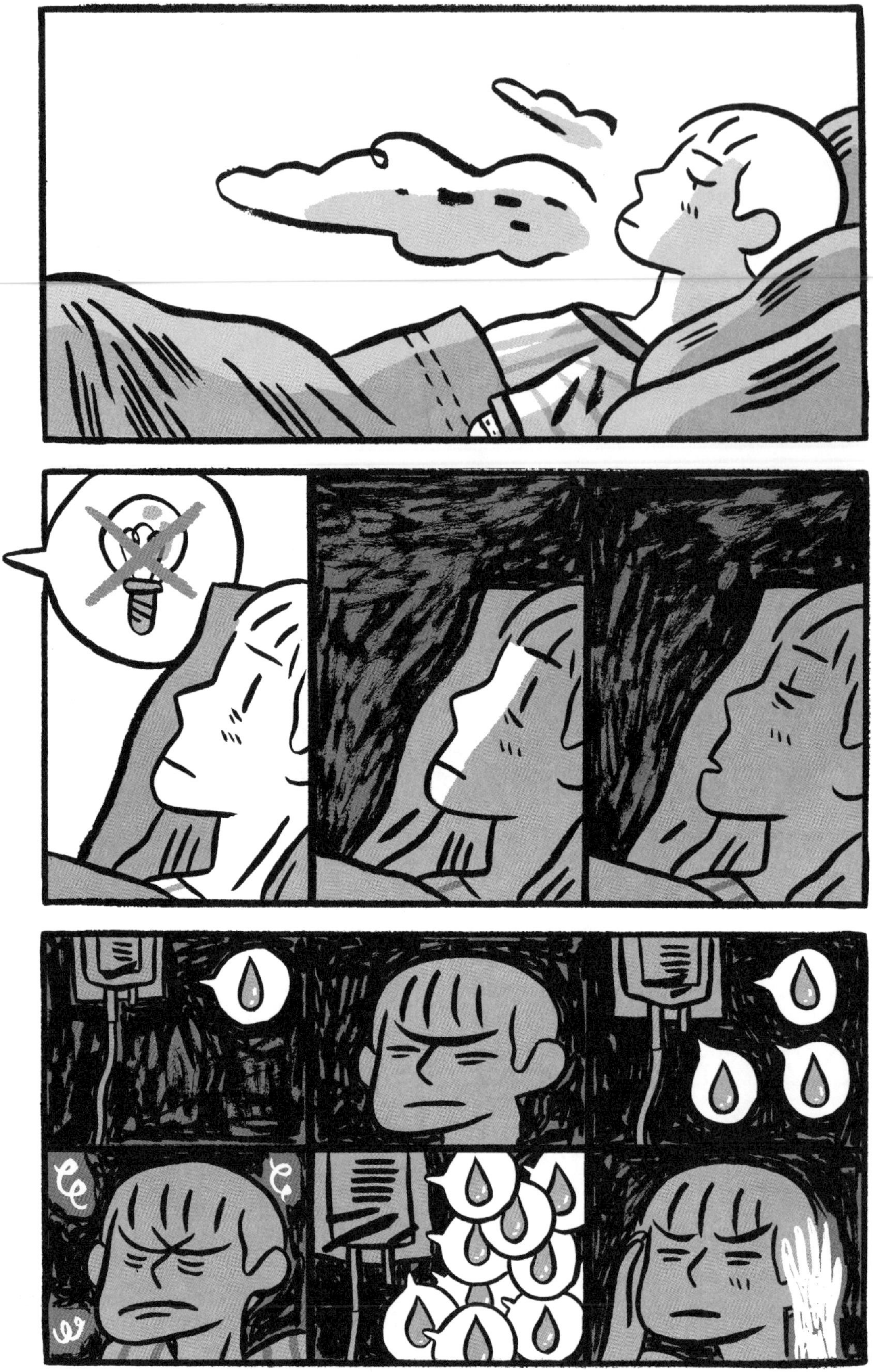

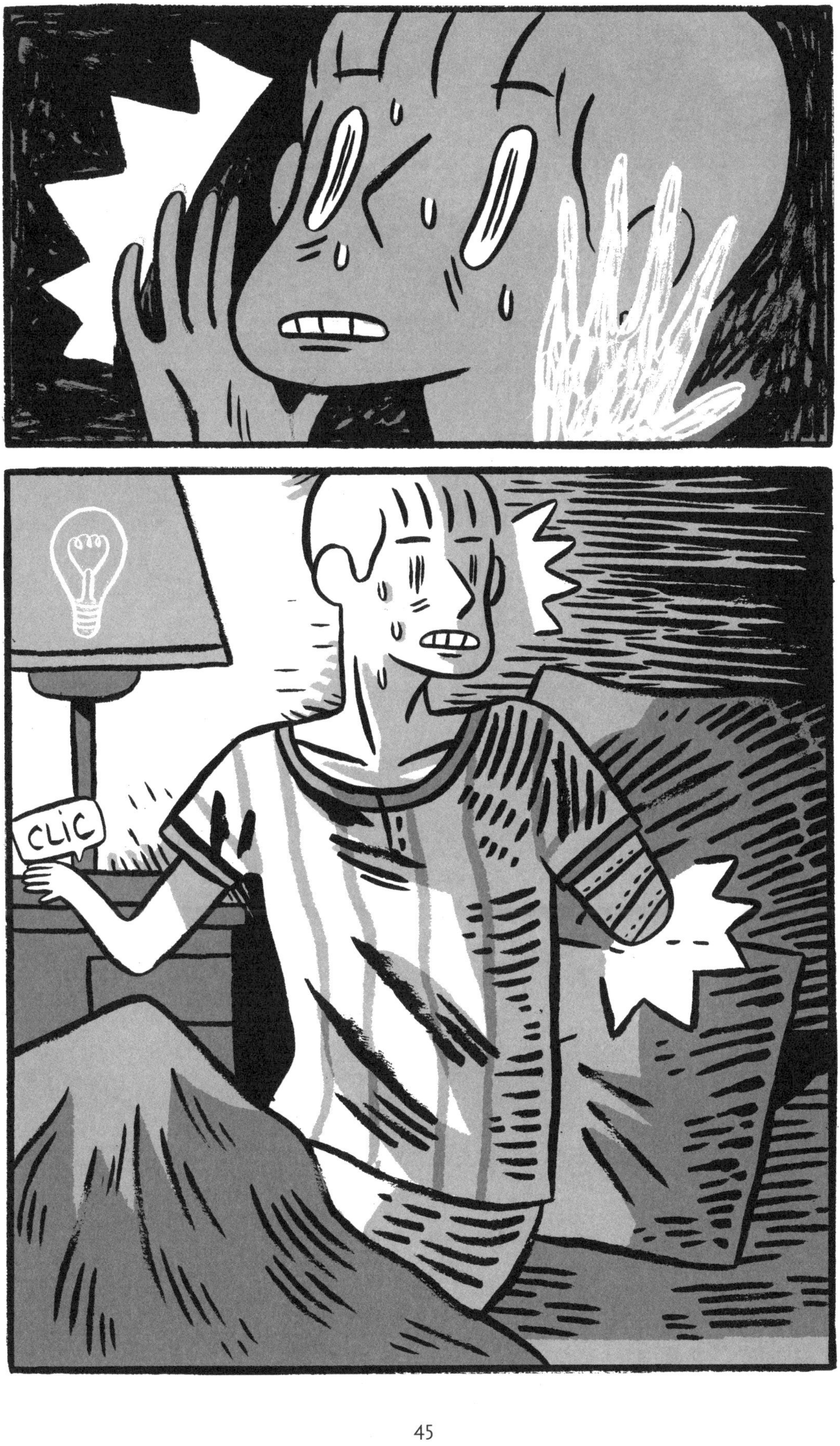
CLIC

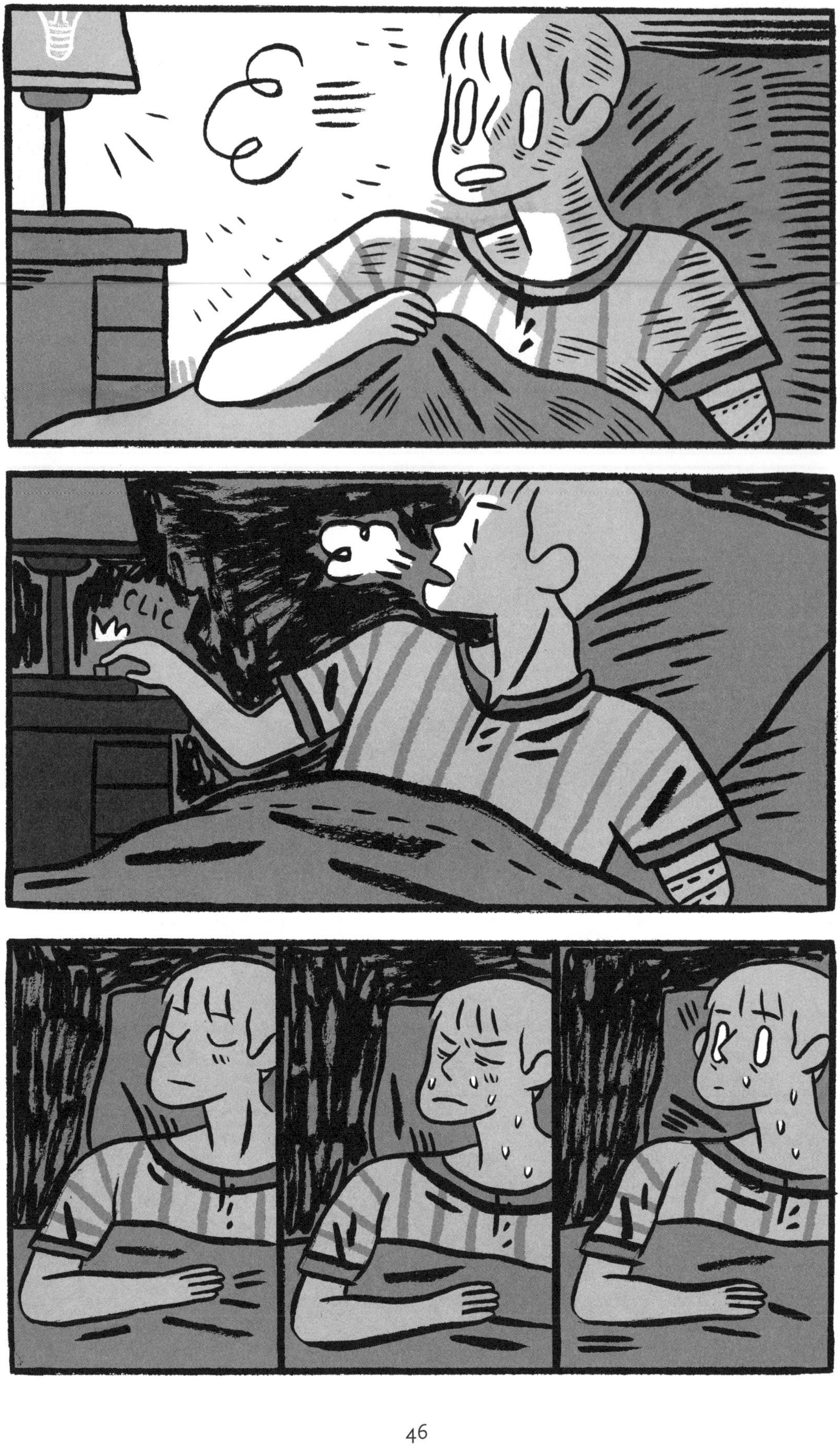
CLIC

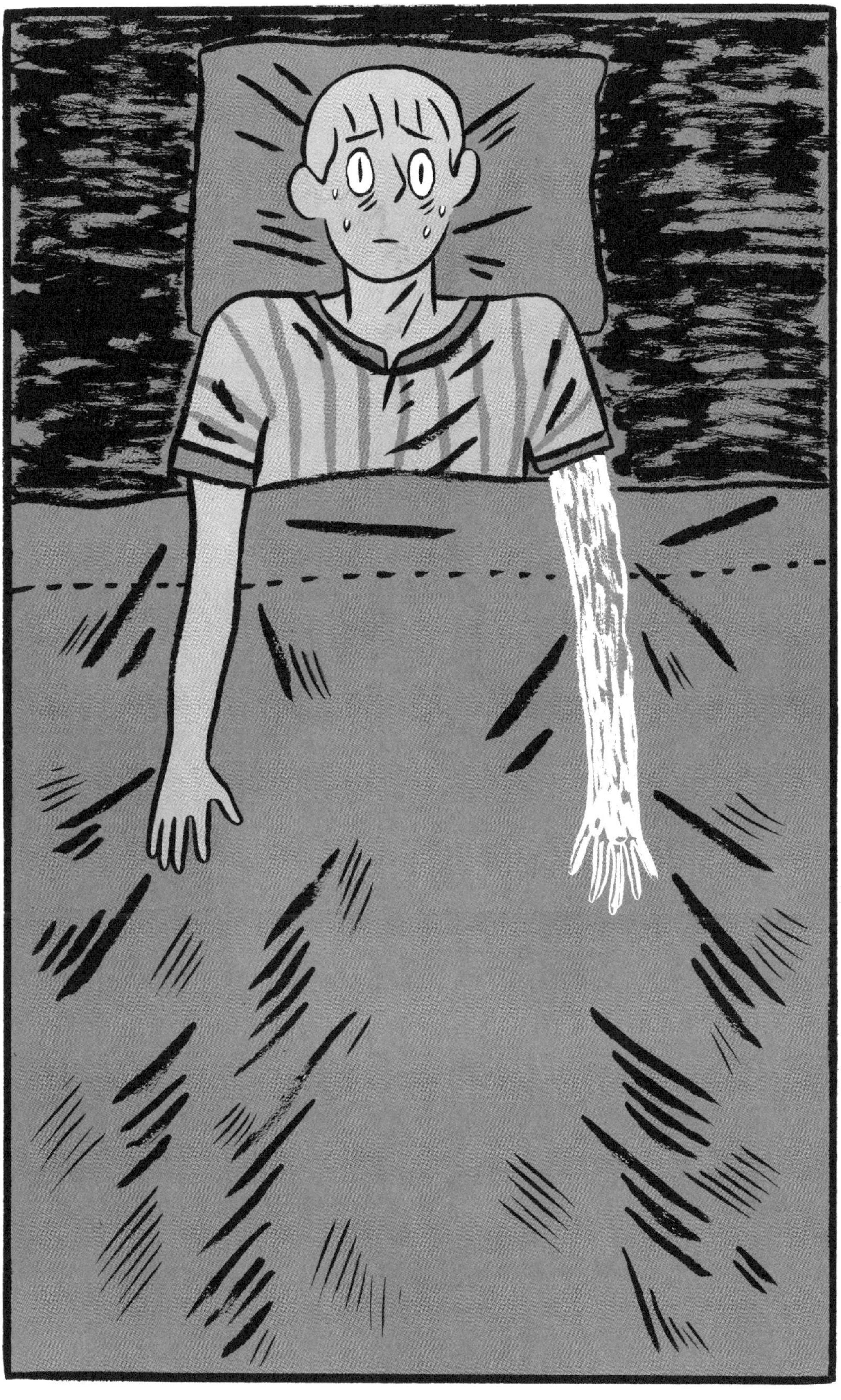

CHAPTER 2
Phantom Limb

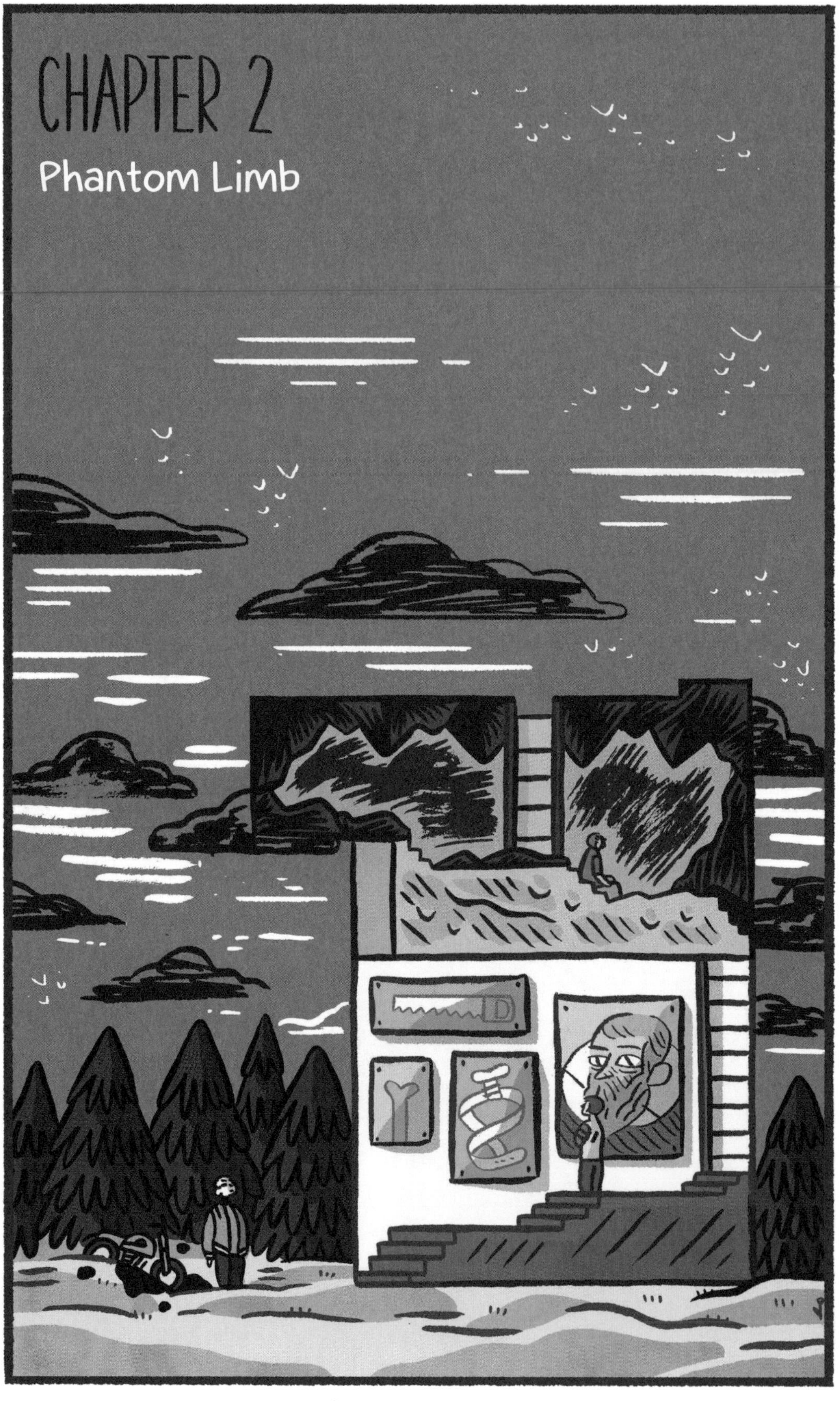

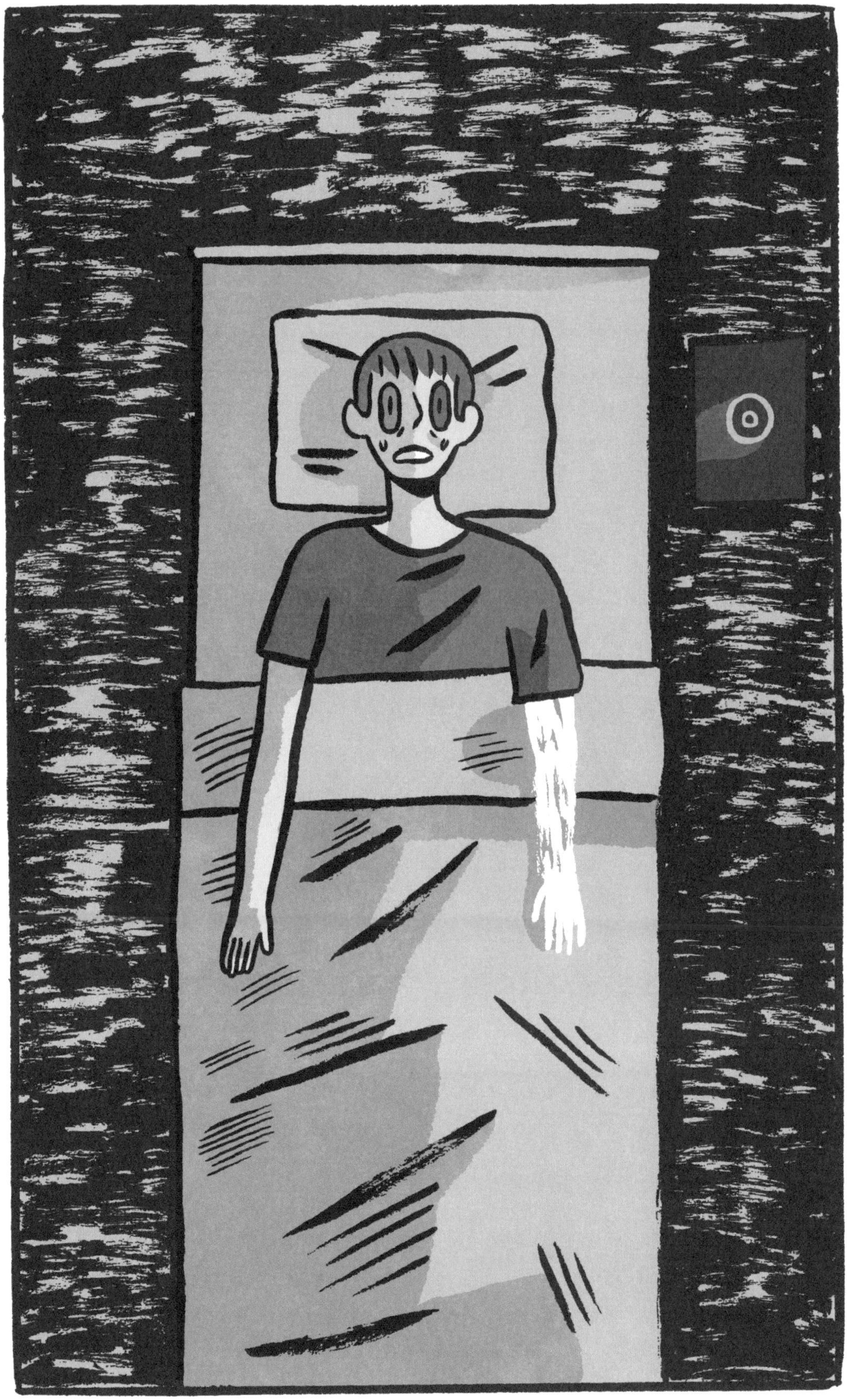

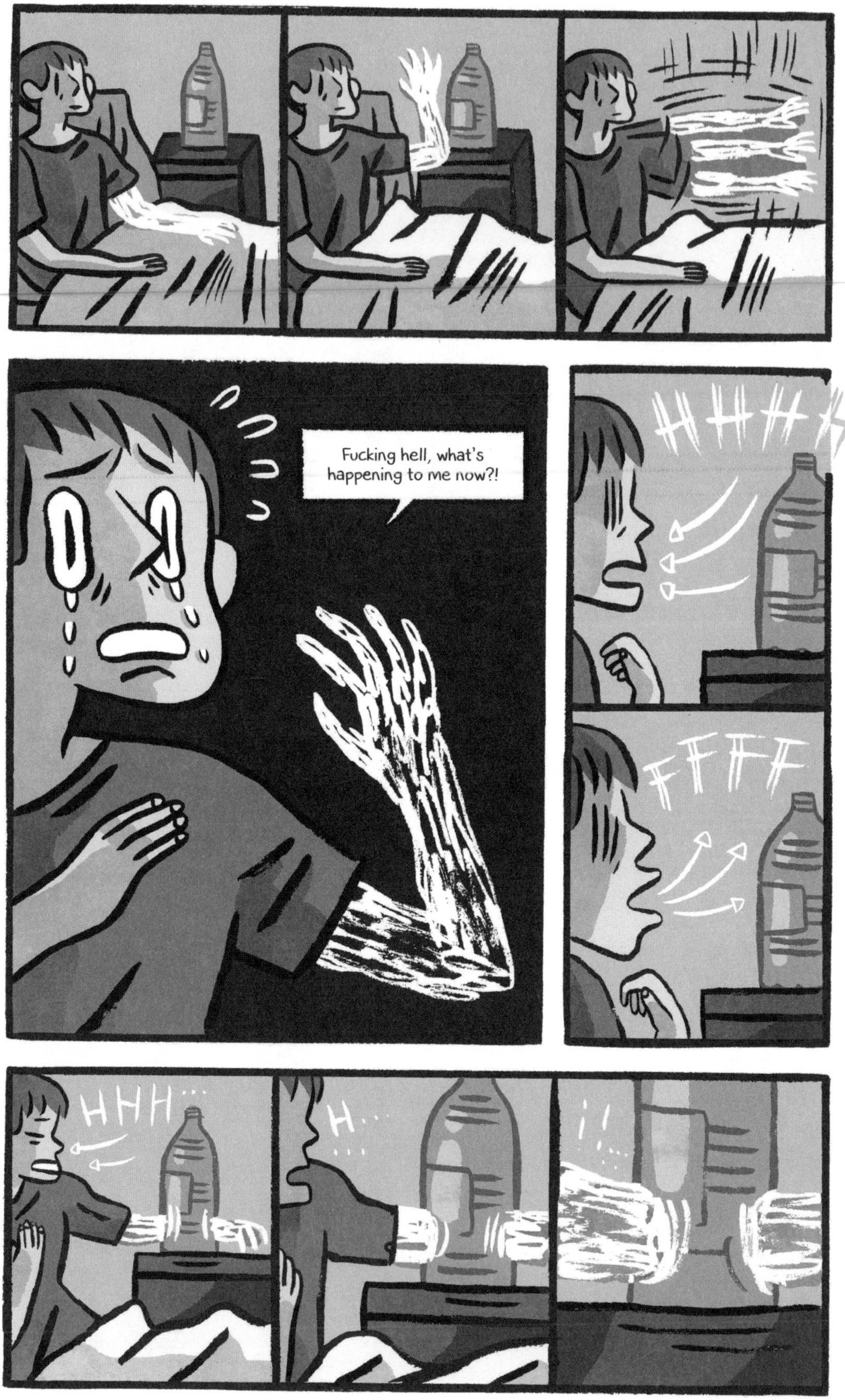
Fucking hell, what's happening to me now?!
HHHH
FFFF
HHH...
H...
!...

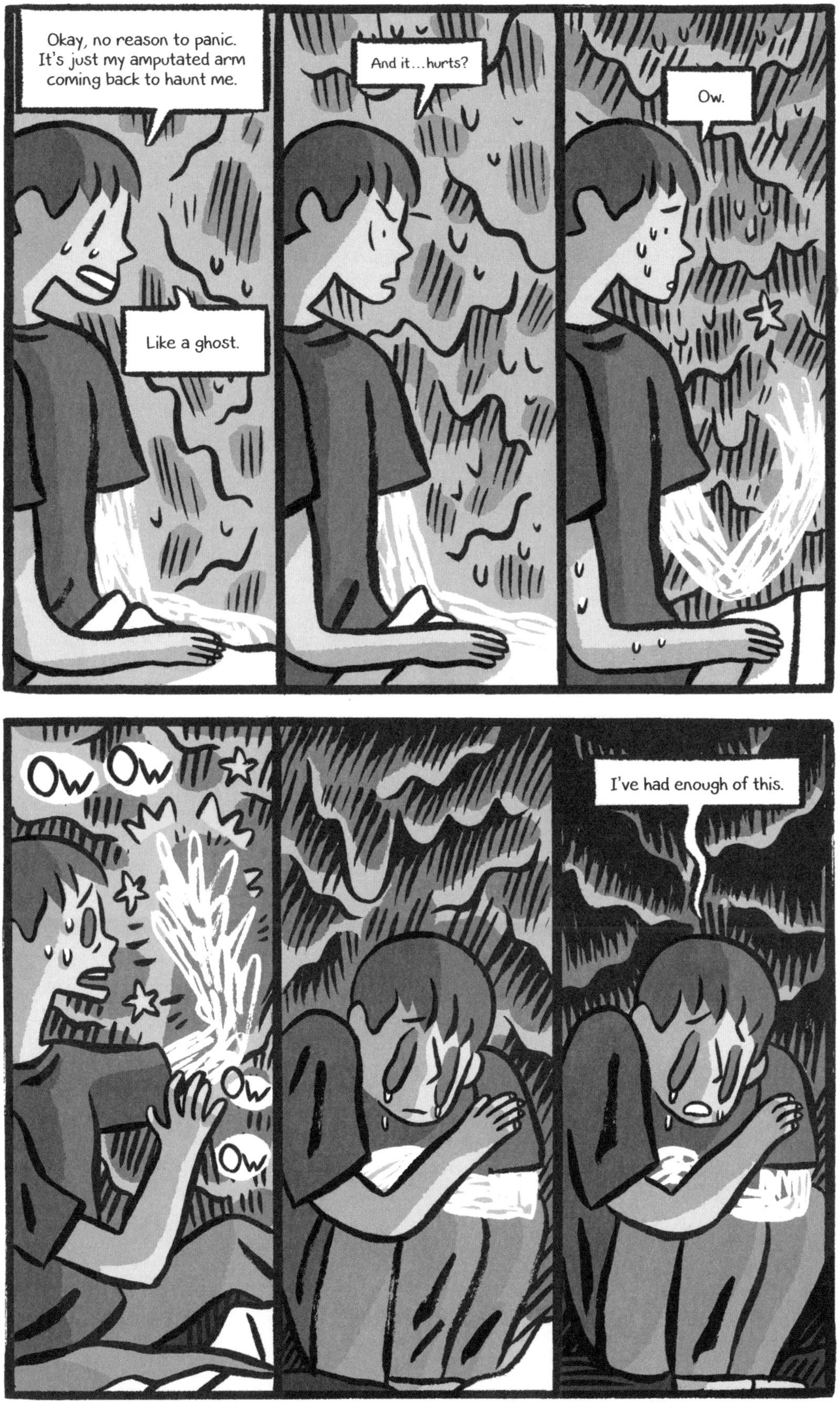
Okay, no reason to panic. It's just my amputated arm coming back to haunt me.
Like a ghost.
And it...hurts?
Ow.
Ow Ow
Ow
Ow
I've had enough of this.

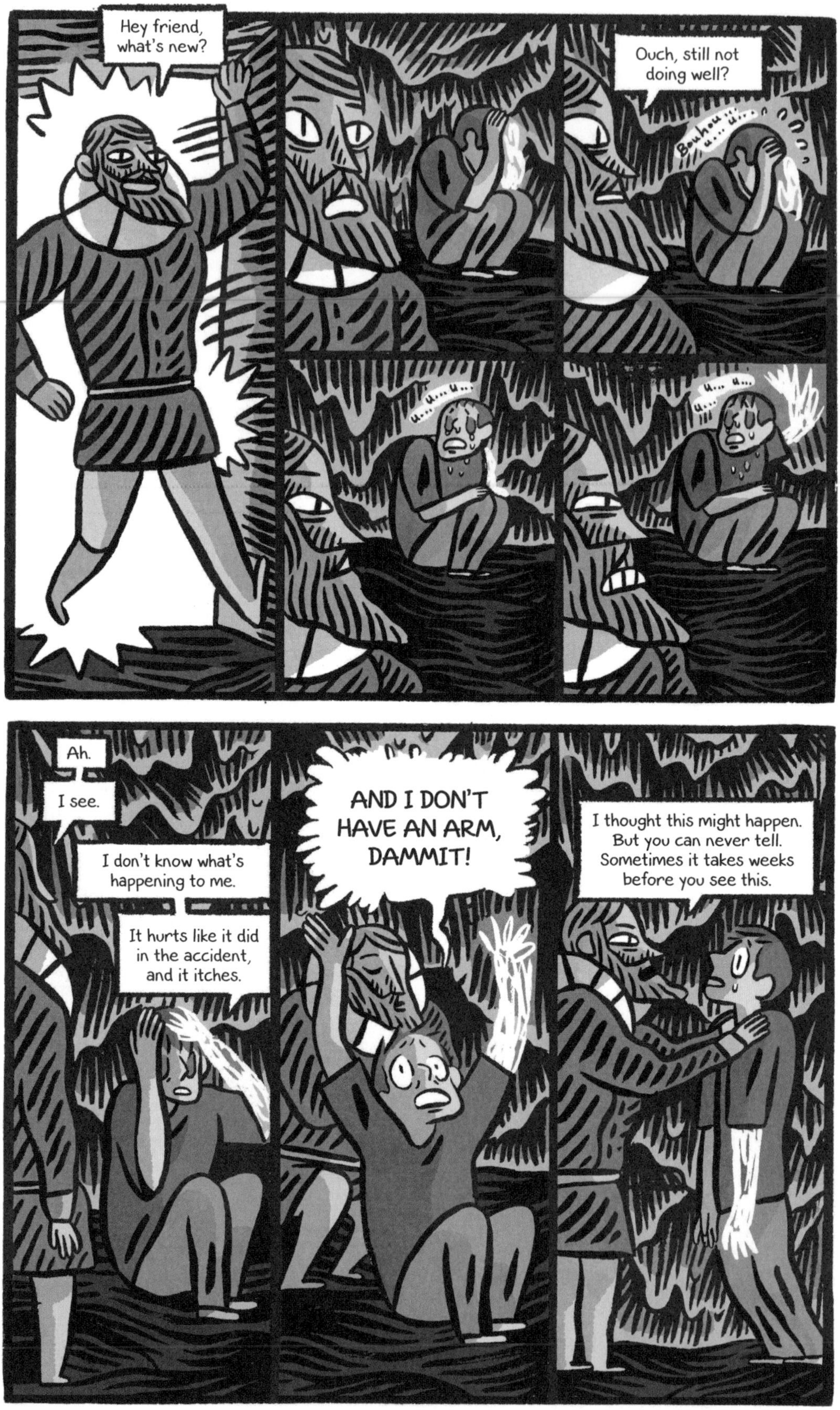
Hey friend, what's new?
Ouch, still not doing well?
Bouhou... u... u...
u... u... u...
u... u... u...
Ah.
I see.
I don't know what's happening to me.
It hurts like it did in the accident, and it itches.
AND I DON'T HAVE AN ARM, DAMMIT!
I thought this might happen. But you can never tell. Sometimes it takes weeks before you see this.

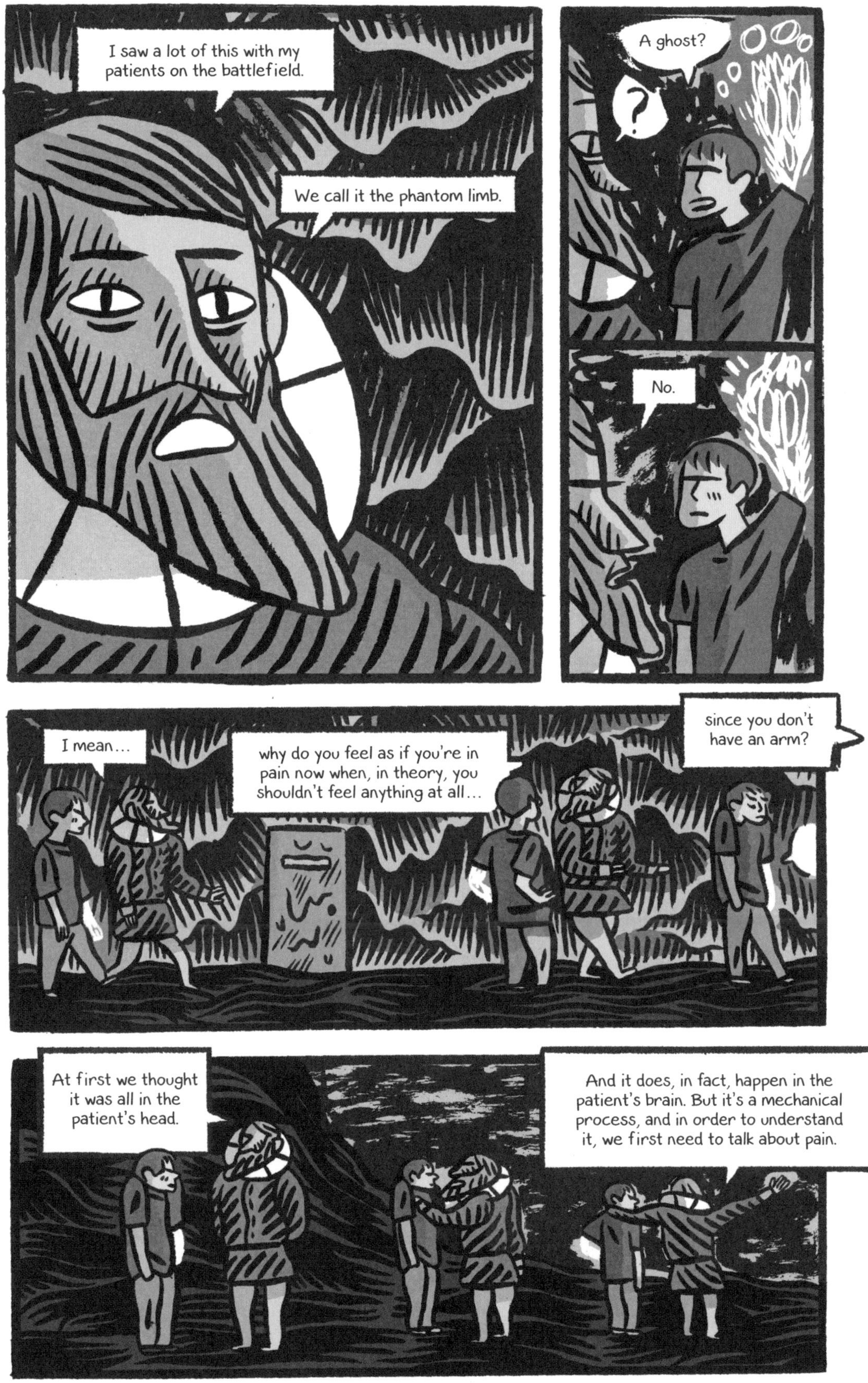
I saw a lot of this with my patients on the battlefield.
We call it the phantom limb.
A ghost?
?
No.
I mean...
why do you feel as if you're in pain now when, in theory, you shouldn't feel anything at all...
since you don't have an arm?
At first we thought it was all in the patient's head.
And it does, in fact, happen in the patient's brain. But it's a mechanical process, and in order to understand it, we first need to talk about pain.

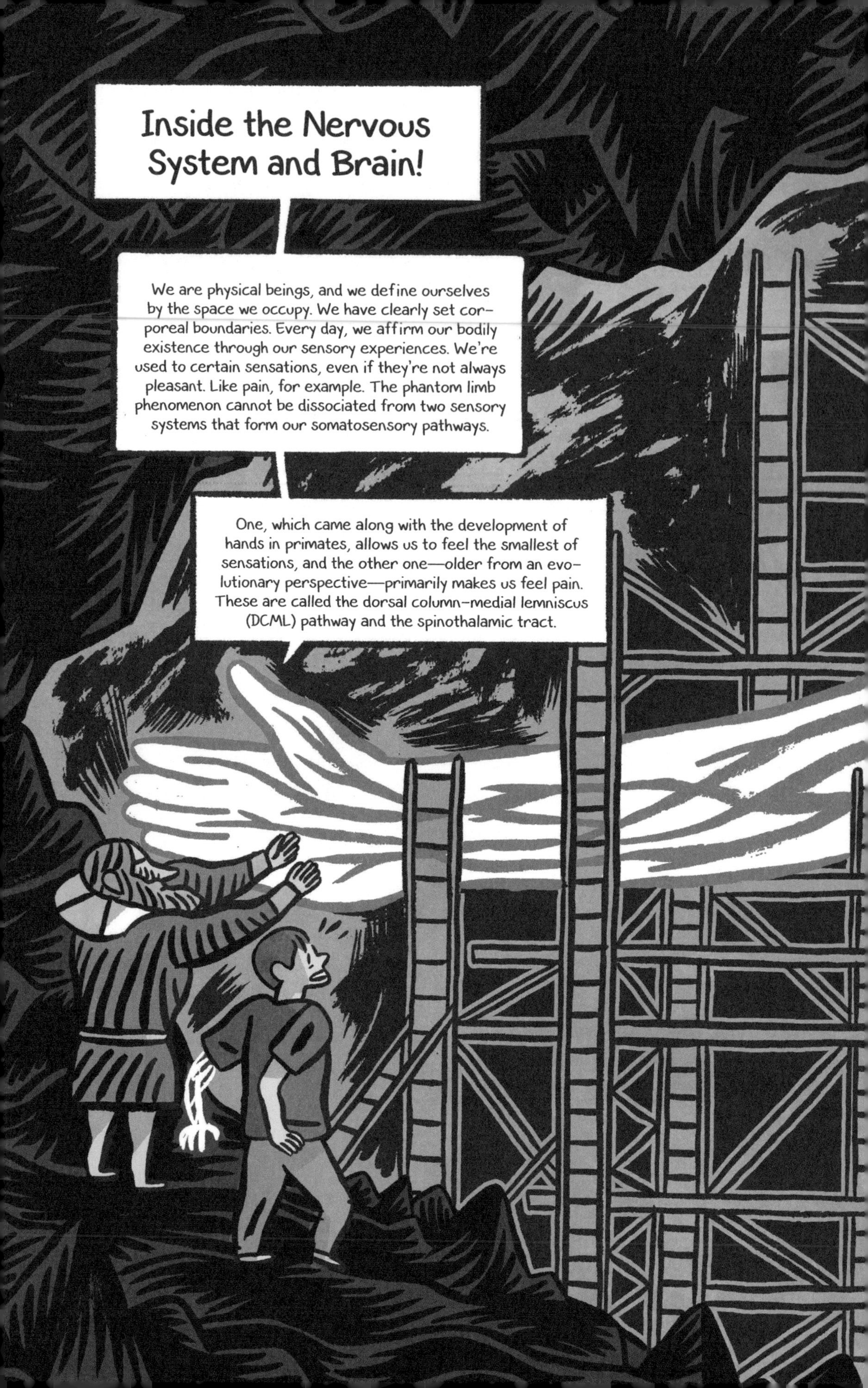
Inside the Nervous System and Brain!
We are physical beings, and we define ourselves by the space we occupy. We have clearly set corporeal boundaries. Every day, we affirm our bodily existence through our sensory experiences. We're used to certain sensations, even if they're not always pleasant. Like pain, for example. The phantom limb phenomenon cannot be dissociated from two sensory systems that form our somatosensory pathways.
One, which came along with the development of hands in primates, allows us to feel the smallest of sensations, and the other one—older from an evolutionary perspective—primarily makes us feel pain. These are called the dorsal column–medial lemniscus (DCML) pathway and the spinothalamic tract.

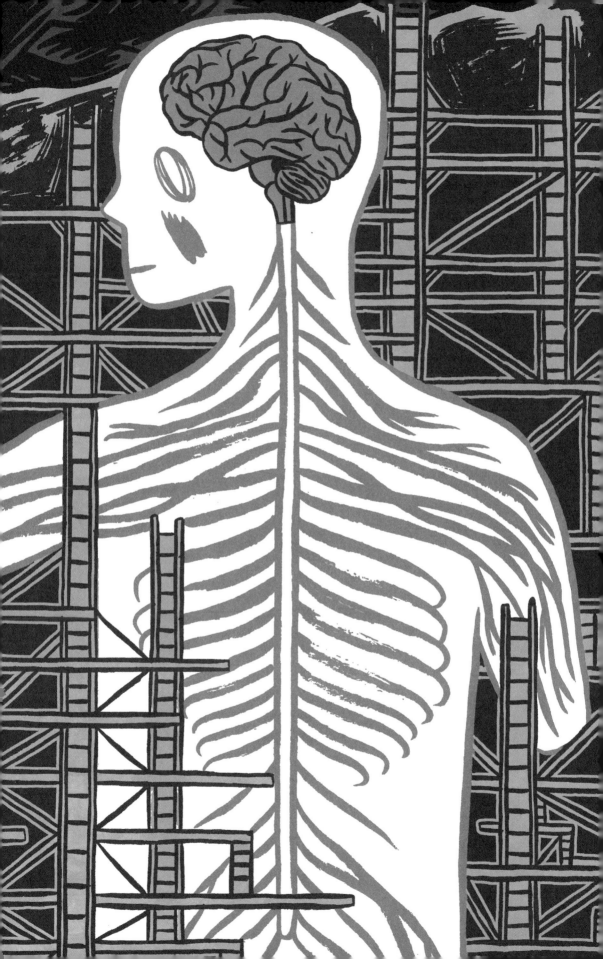

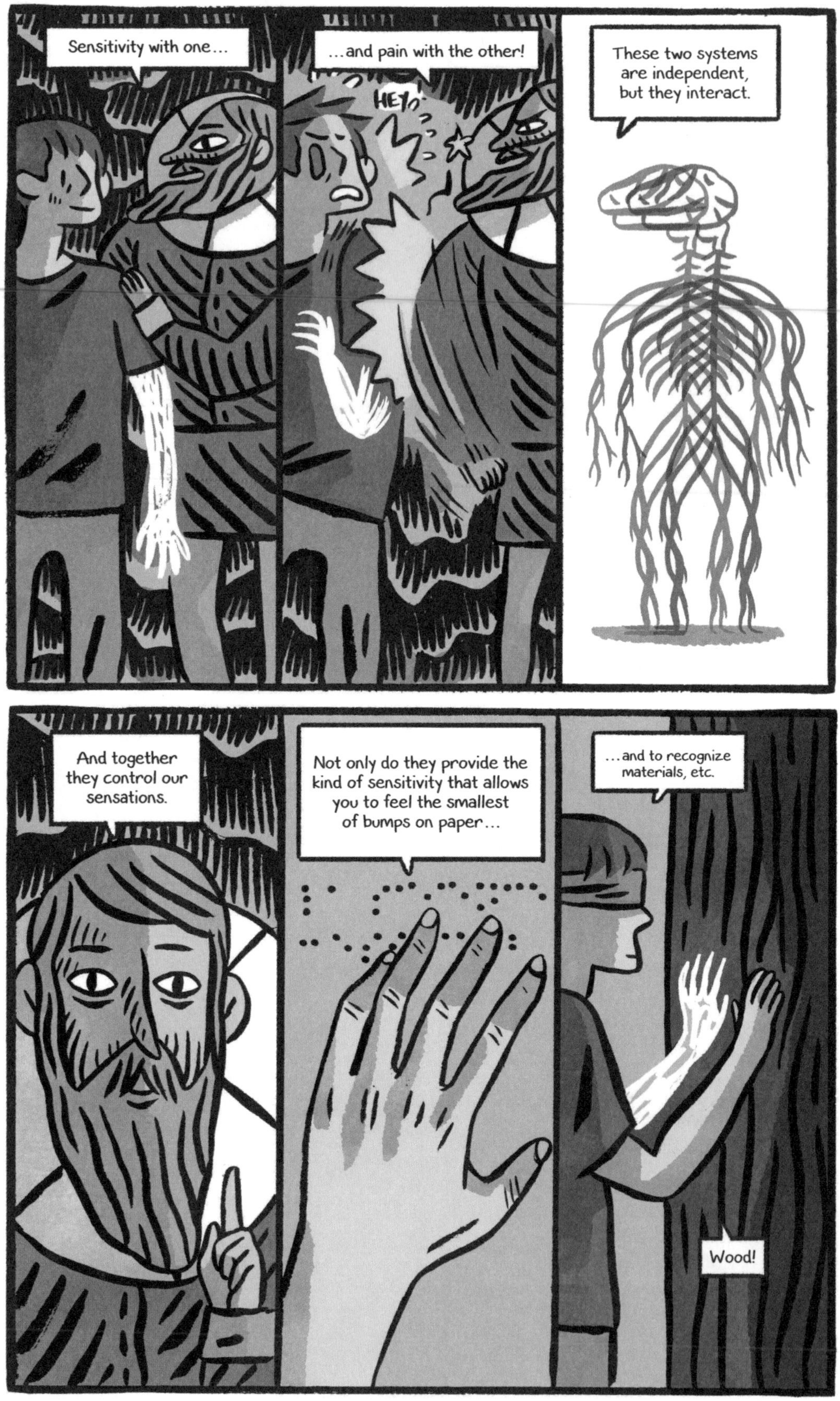

Sensitivity with one...
...and pain with the other!
HEY!
These two systems are independent, but they interact.
And together they control our sensations.
Not only do they provide the kind of sensitivity that allows you to feel the smallest of bumps on paper...
...and to recognize materials, etc.
Wood!

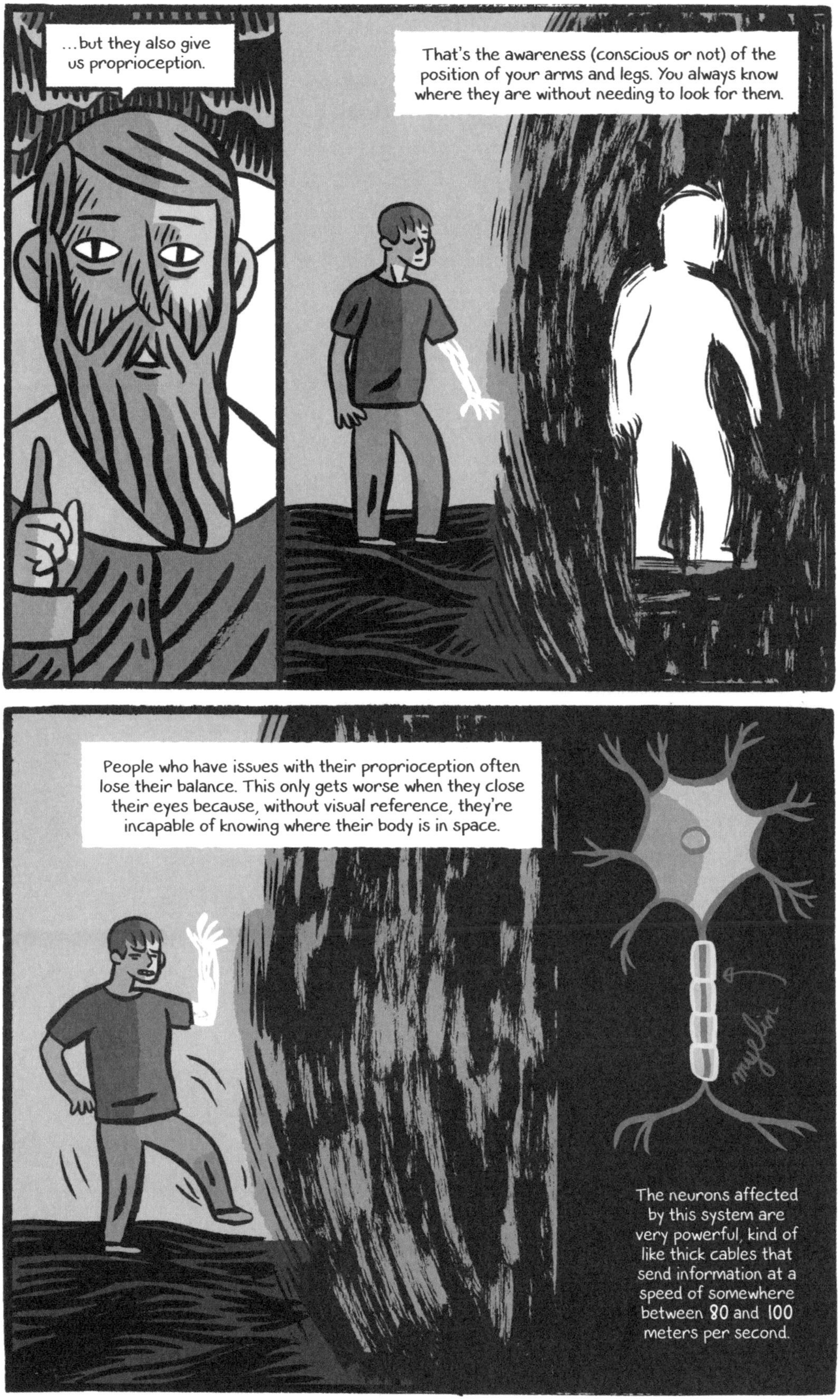
...but they also give us proprioception.
That's the awareness (conscious or not) of the position of your arms and legs. You always know where they are without needing to look for them.
People who have issues with their proprioception often lose their balance. This only gets worse when they close their eyes because, without visual reference, they're incapable of knowing where their body is in space.
myelin
The neurons affected by this system are very powerful, kind of like thick cables that send information at a speed of somewhere between 80 and 100 meters per second.

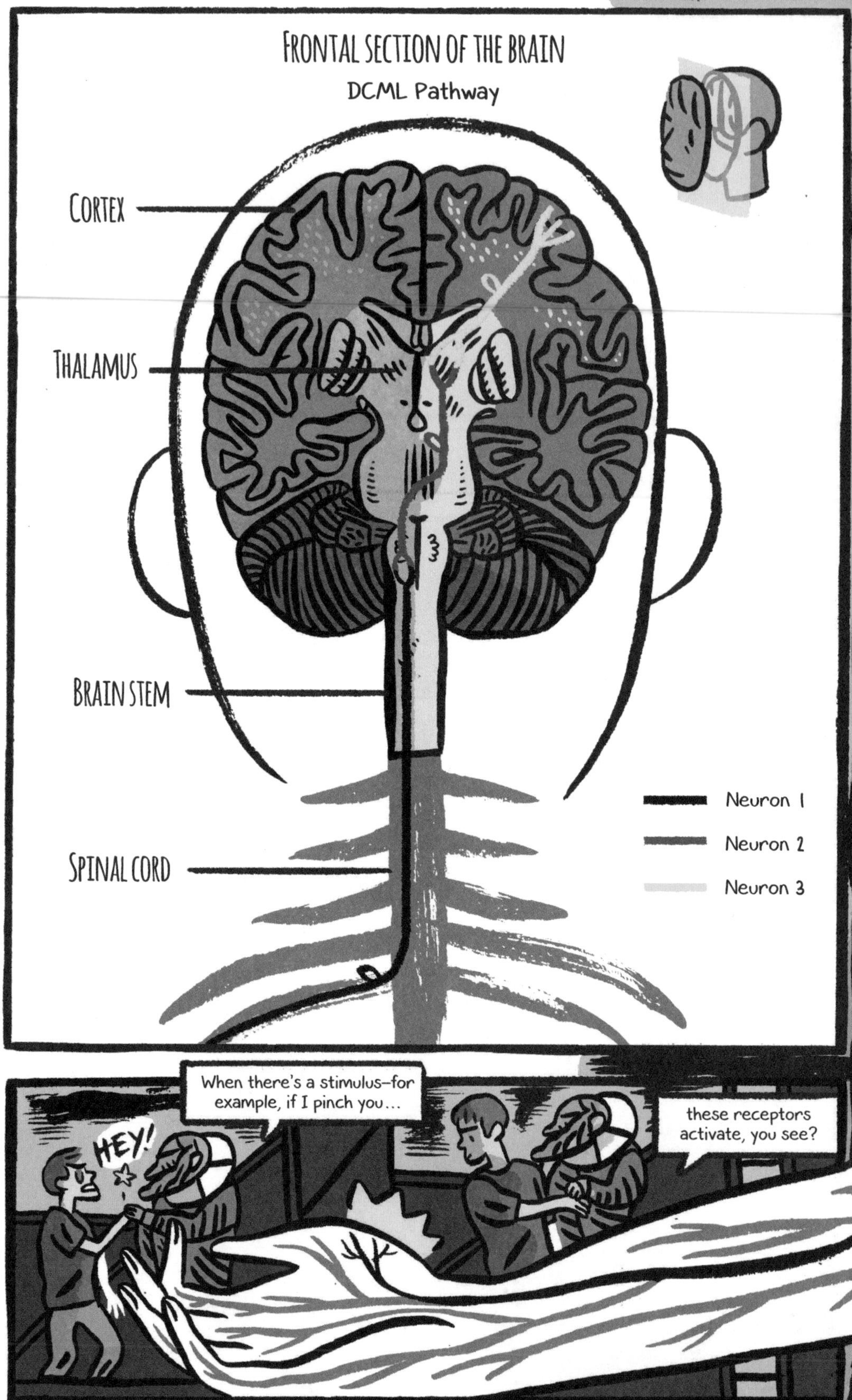

FRONTAL SECTION OF THE BRAIN
DCML Pathway
CORTEX
THALAMUS
BRAIN STEM
SPINAL CORD
Neuron 1
Neuron 2
Neuron 3
When there's a stimulus—for example, if I pinch you...
HEY!
these receptors activate, you see?

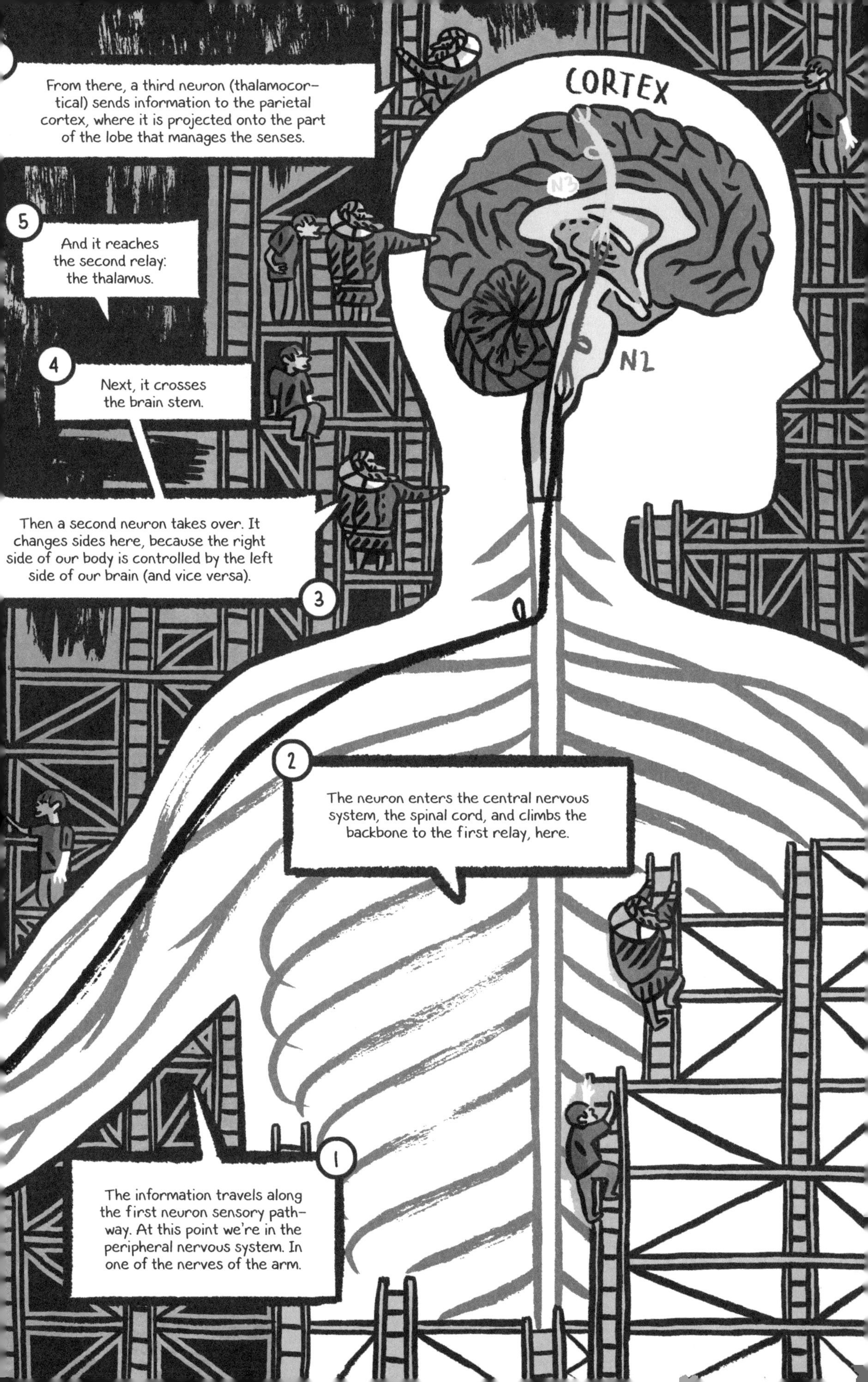

From there, a third neuron (thalamocortical) sends information to the parietal cortex, where it is projected onto the part of the lobe that manages the senses.
CORTEX
N3
5
And it reaches the second relay: the thalamus.
N2
4
Next, it crosses the brain stem.
Then a second neuron takes over. It changes sides here, because the right side of our body is controlled by the left side of our brain (and vice versa).
3
2
The neuron enters the central nervous system, the spinal cord, and climbs the backbone to the first relay, here.
1
The information travels along the first neuron sensory pathway. At this point we're in the peripheral nervous system. In one of the nerves of the arm.

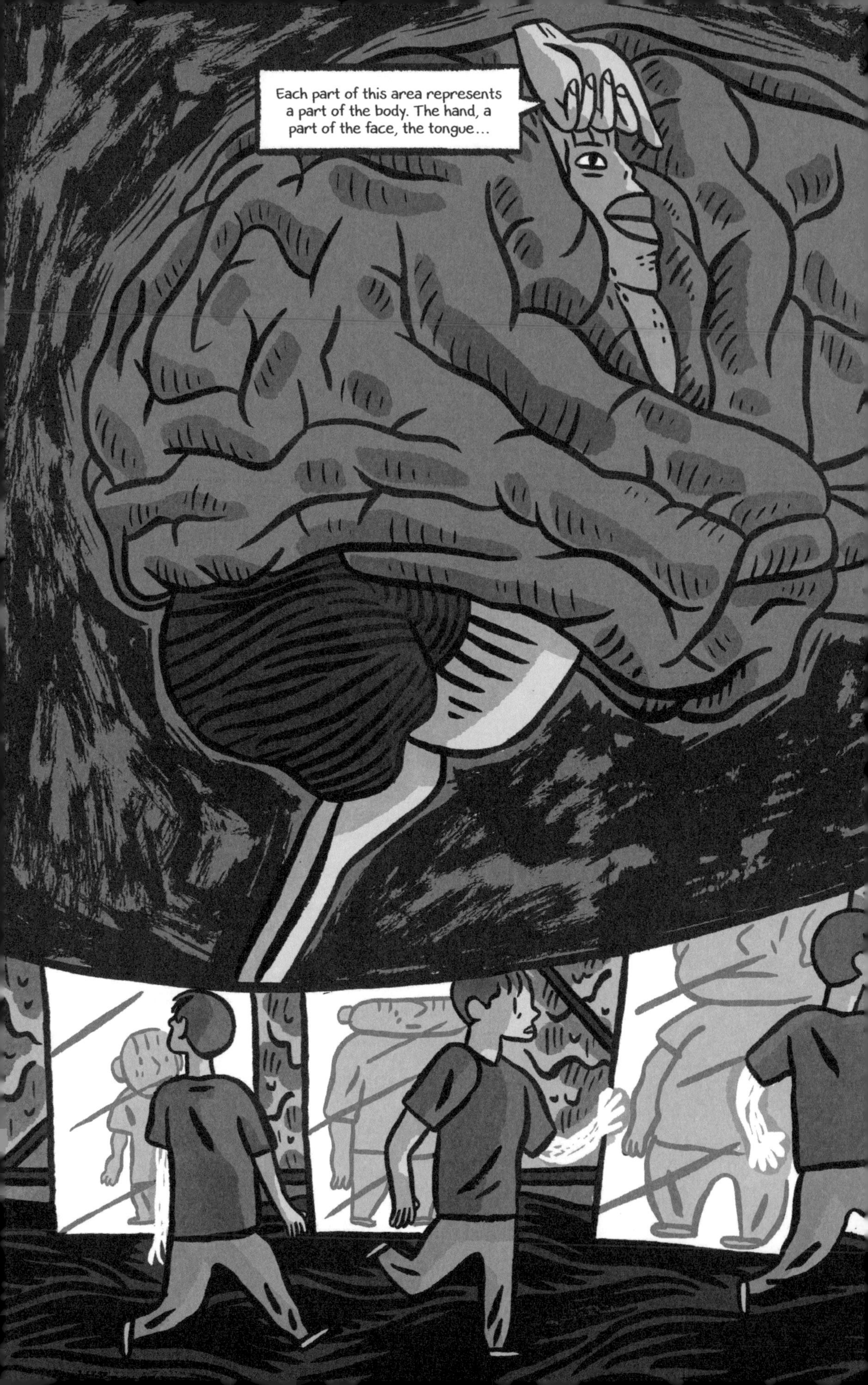
Each part of this area represents a part of the body. The hand, a part of the face, the tongue...

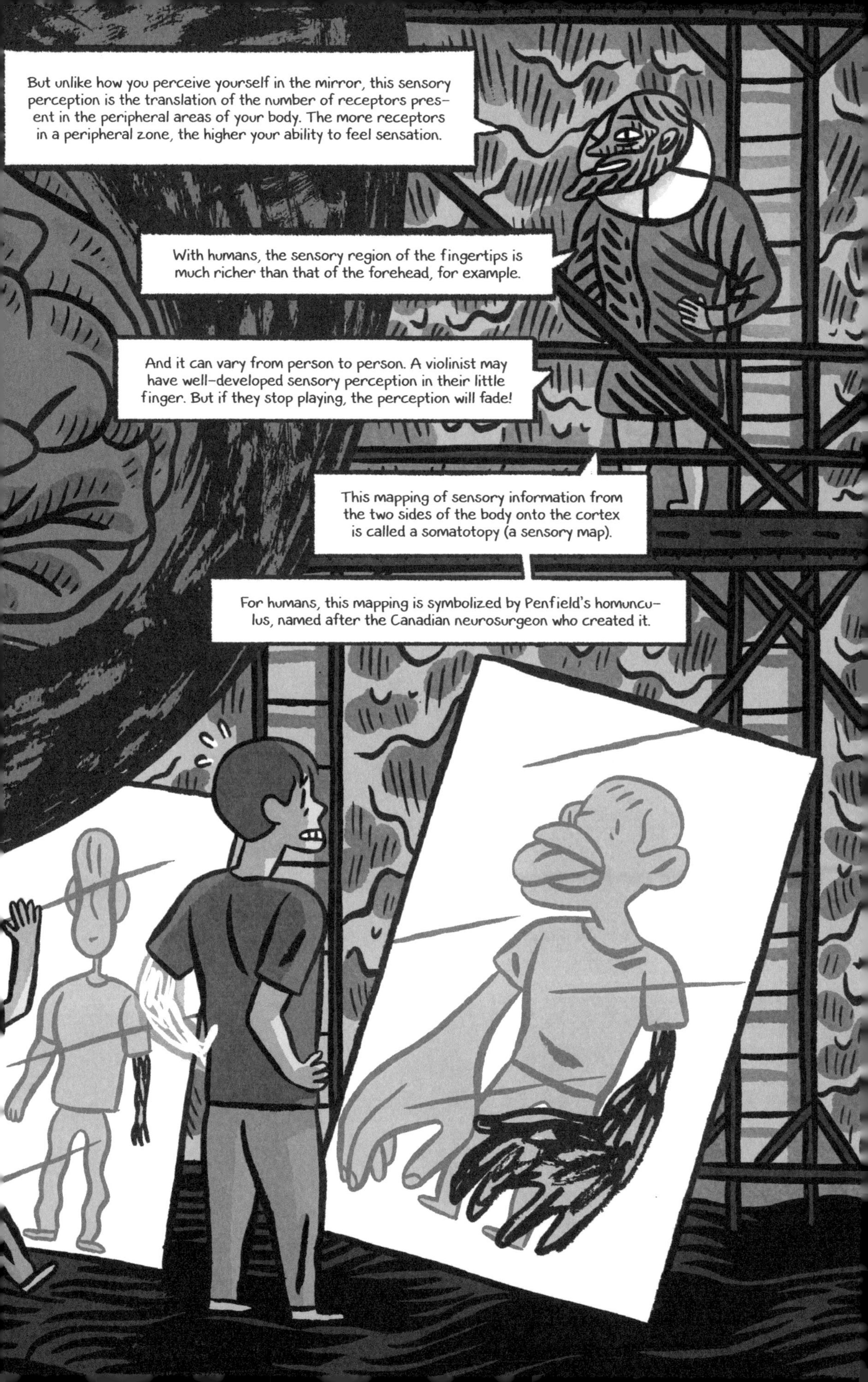
But unlike how you perceive yourself in the mirror, this sensory perception is the translation of the number of receptors present in the peripheral areas of your body. The more receptors in a peripheral zone, the higher your ability to feel sensation.
With humans, the sensory region of the fingertips is much richer than that of the forehead, for example.
And it can vary from person to person. A violinist may have well-developed sensory perception in their little finger. But if they stop playing, the perception will fade!
This mapping of sensory information from the two sides of the body onto the cortex is called a somatotopy (a sensory map).
For humans, this mapping is symbolized by Penfield's homunculus, named after the Canadian neurosurgeon who created it.

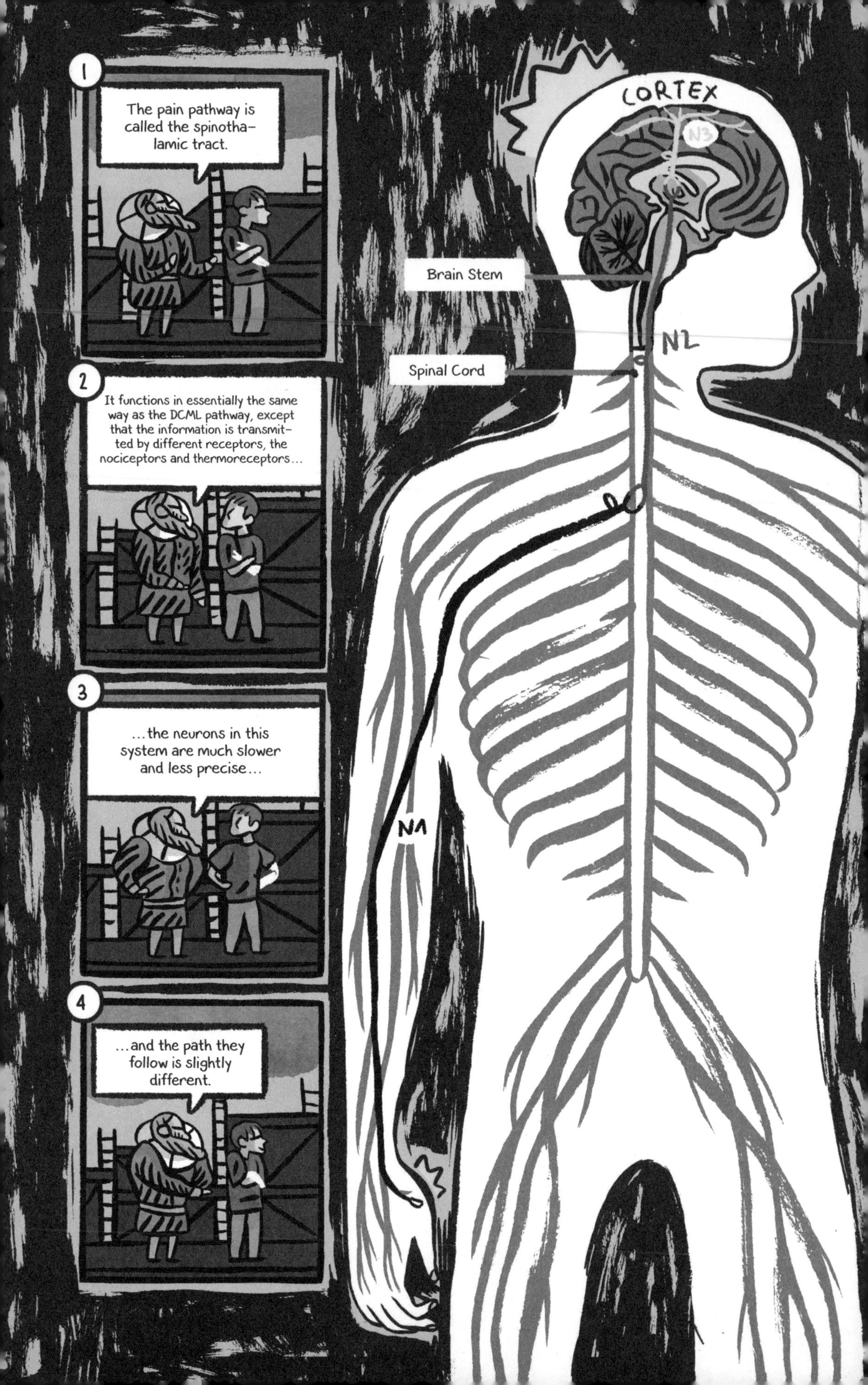

1
The pain pathway is called the spinothalamic tract.
2
It functions in essentially the same way as the DCML pathway, except that the information is transmitted by different receptors, the nociceptors and thermoreceptors...
3
...the neurons in this system are much slower and less precise...
4
...and the path they follow is slightly different.
CORTEX
N3
Brain Stem
N2
Spinal Cord
N1

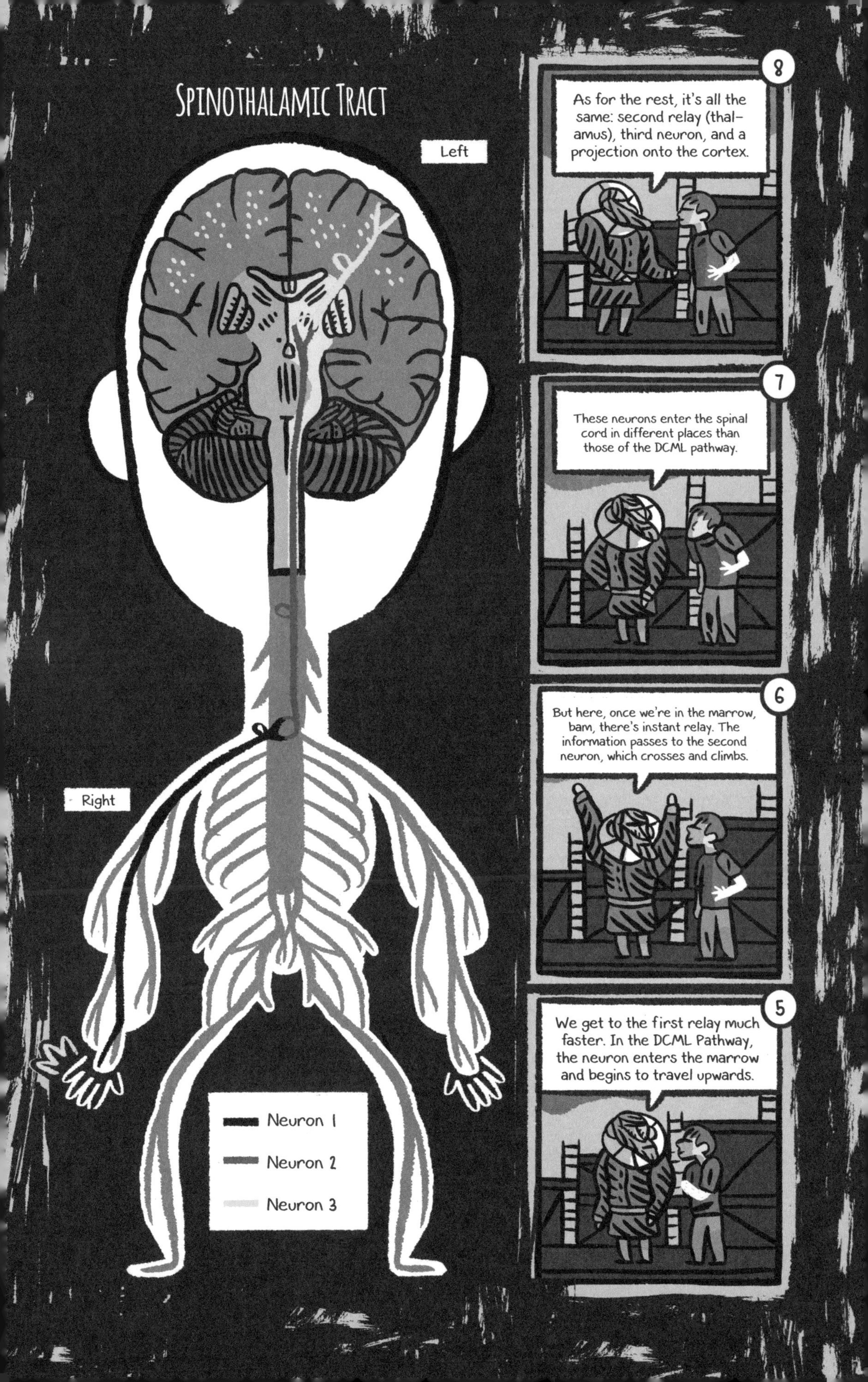

Spinothalamic Tract
Left
Right
Neuron 1
Neuron 2
Neuron 3
5
We get to the first relay much faster. In the DCML Pathway, the neuron enters the marrow and begins to travel upwards.
6
But here, once we're in the marrow, bam, there's instant relay. The information passes to the second neuron, which crosses and climbs.
7
These neurons enter the spinal cord in different places than those of the DCML pathway.
8
As for the rest, it's all the same: second relay (thal-amus), third neuron, and a projection onto the cortex.

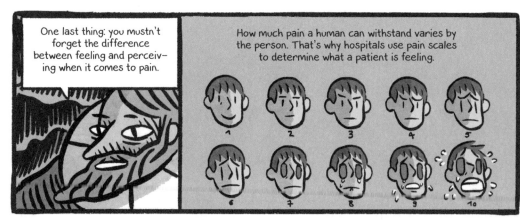

One last thing: you mustn't forget the difference between feeling and perceiving when it comes to pain.

How much pain a human can withstand varies by the person. That's why hospitals use pain scales to determine what a patient is feeling.

Without the spinothalamic tract, we wouldn't feel pain.

Yes, that's true.

A life without pain... sounds kinda nice.

Think so? Not really.

Pain is necessary! It teaches us to avoid dangerous situations and keep ourselves safe. Imagine accidentally putting your hand on an electric stove. If you didn't feel the pain, you could get cooked.

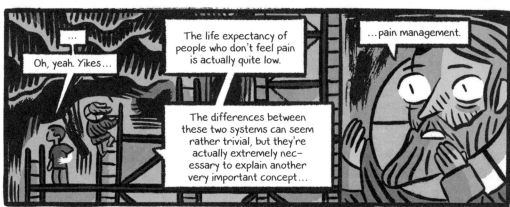

...

Oh, yeah. Yikes...

The life expectancy of people who don't feel pain is actually quite low.

...pain management.

The differences between these two systems can seem rather trivial, but they're actually extremely necessary to explain another very important concept...

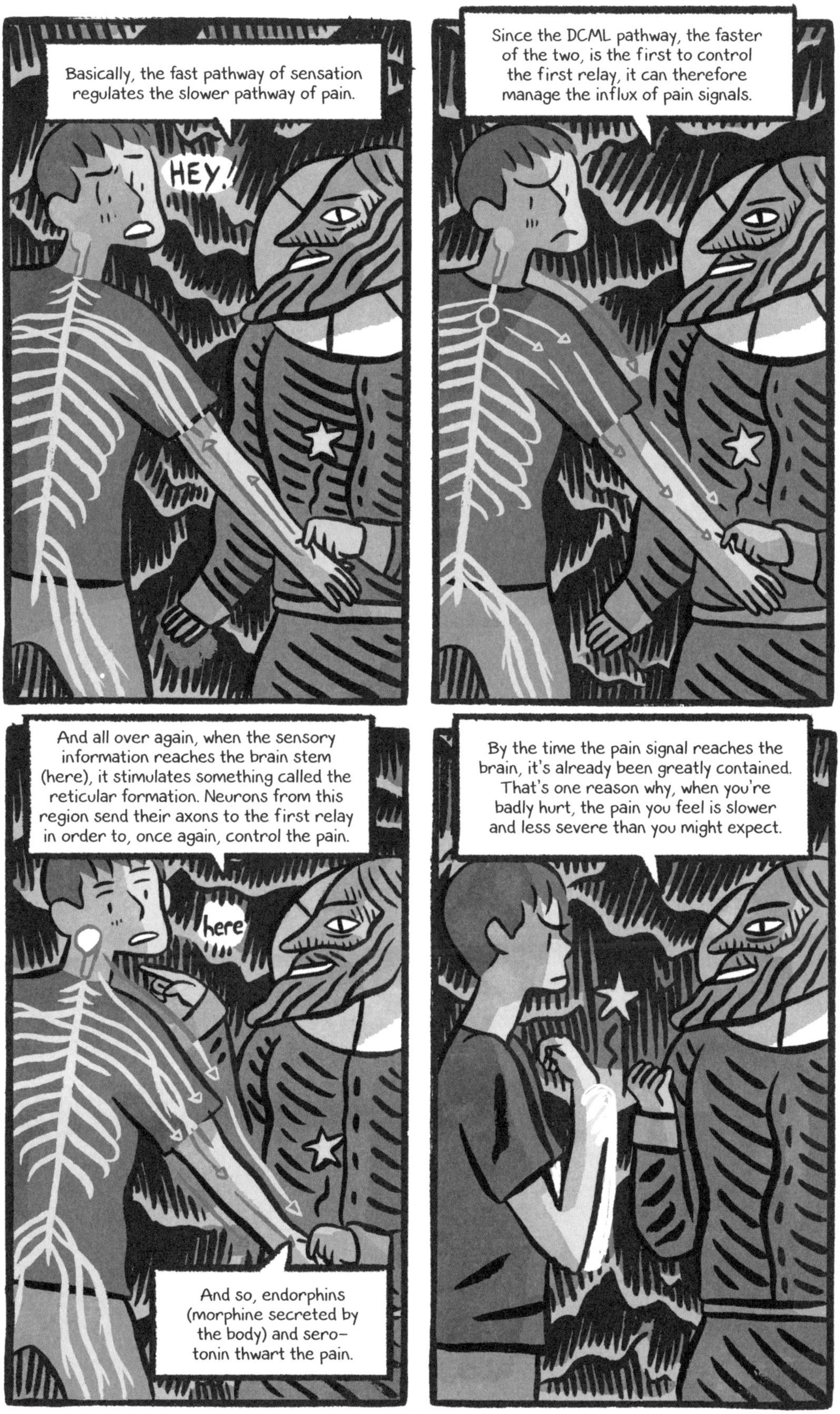
Basically, the fast pathway of sensation regulates the slower pathway of pain.
HEY!
Since the DCML pathway, the faster of the two, is the first to control the first relay, it can therefore manage the influx of pain signals.
And all over again, when the sensory information reaches the brain stem (here), it stimulates something called the reticular formation. Neurons from this region send their axons to the first relay in order to, once again, control the pain.
here
And so, endorphins (morphine secreted by the body) and serotonin thwart the pain.
By the time the pain signal reaches the brain, it's already been greatly contained. That's one reason why, when you're badly hurt, the pain you feel is slower and less severe than you might expect.

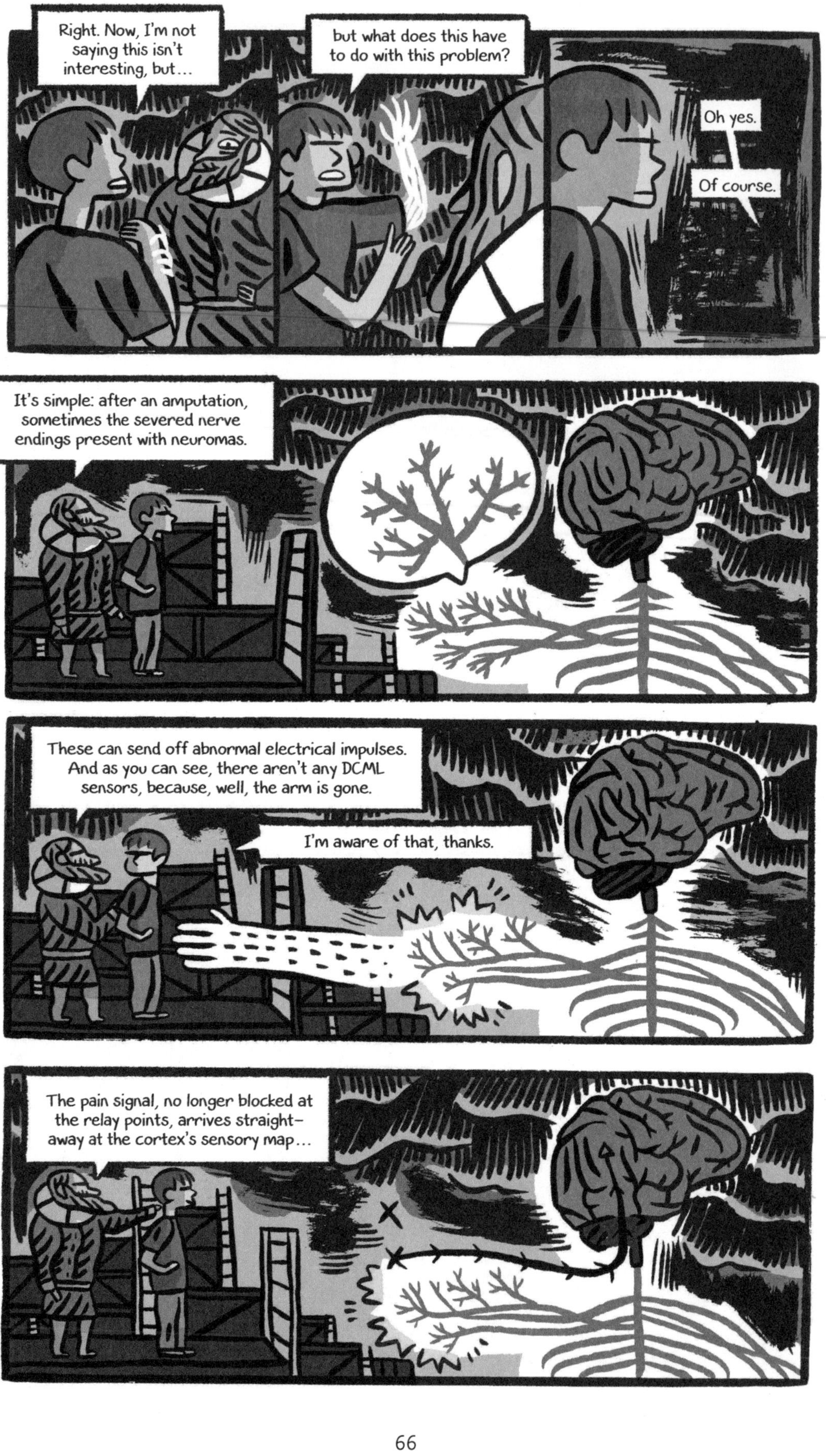
Right. Now, I'm not saying this isn't interesting, but...
but what does this have to do with this problem?
Oh yes.
Of course.
It's simple: after an amputation, sometimes the severed nerve endings present with neuromas.
These can send off abnormal electrical impulses. And as you can see, there aren't any DCML sensors, because, well, the arm is gone.
I'm aware of that, thanks.
The pain signal, no longer blocked at the relay points, arrives straight-away at the cortex's sensory map...

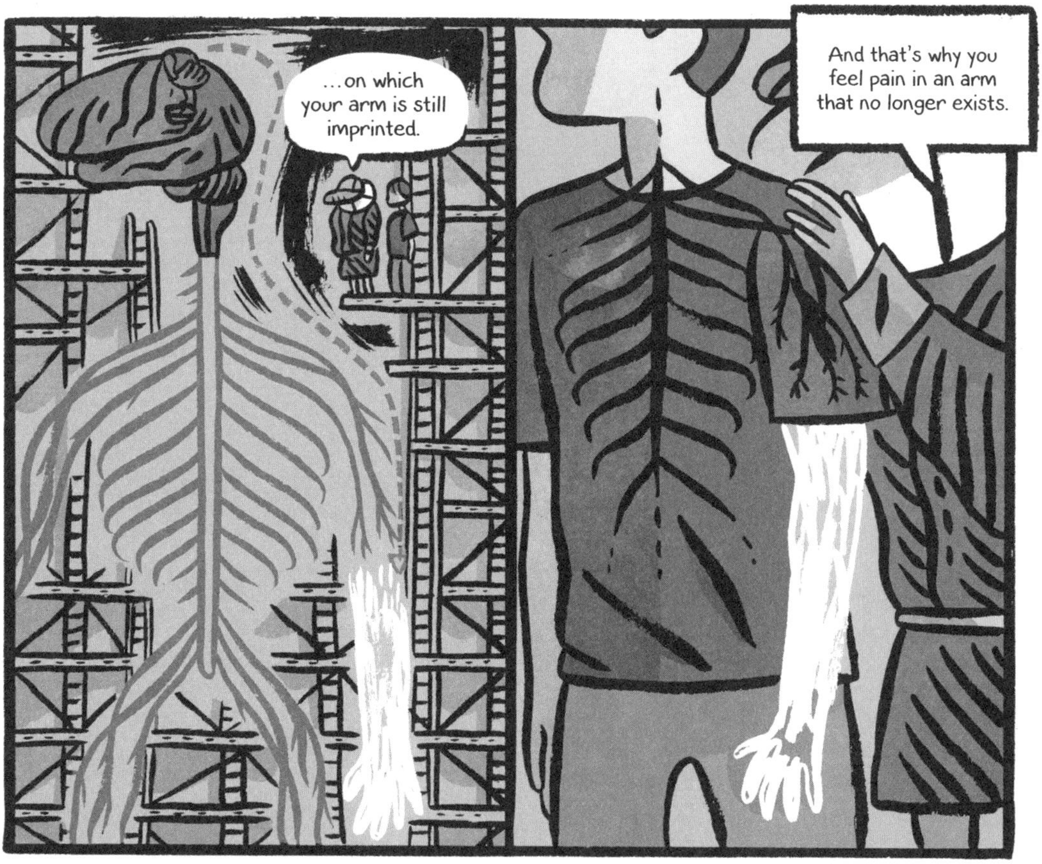
...on which your arm is still imprinted.
And that's why you feel pain in an arm that no longer exists.

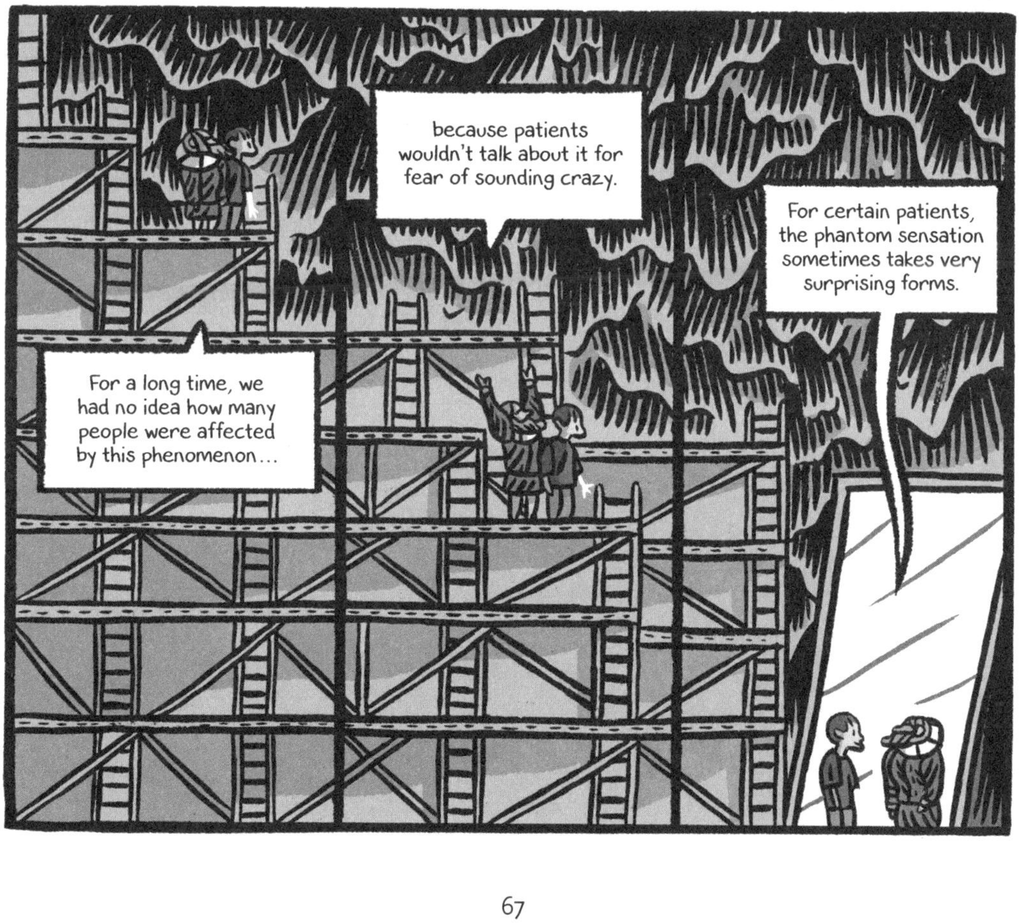
For a long time, we had no idea how many people were affected by this phenomenon...
because patients wouldn't talk about it for fear of sounding crazy.
For certain patients, the phantom sensation sometimes takes very surprising forms.

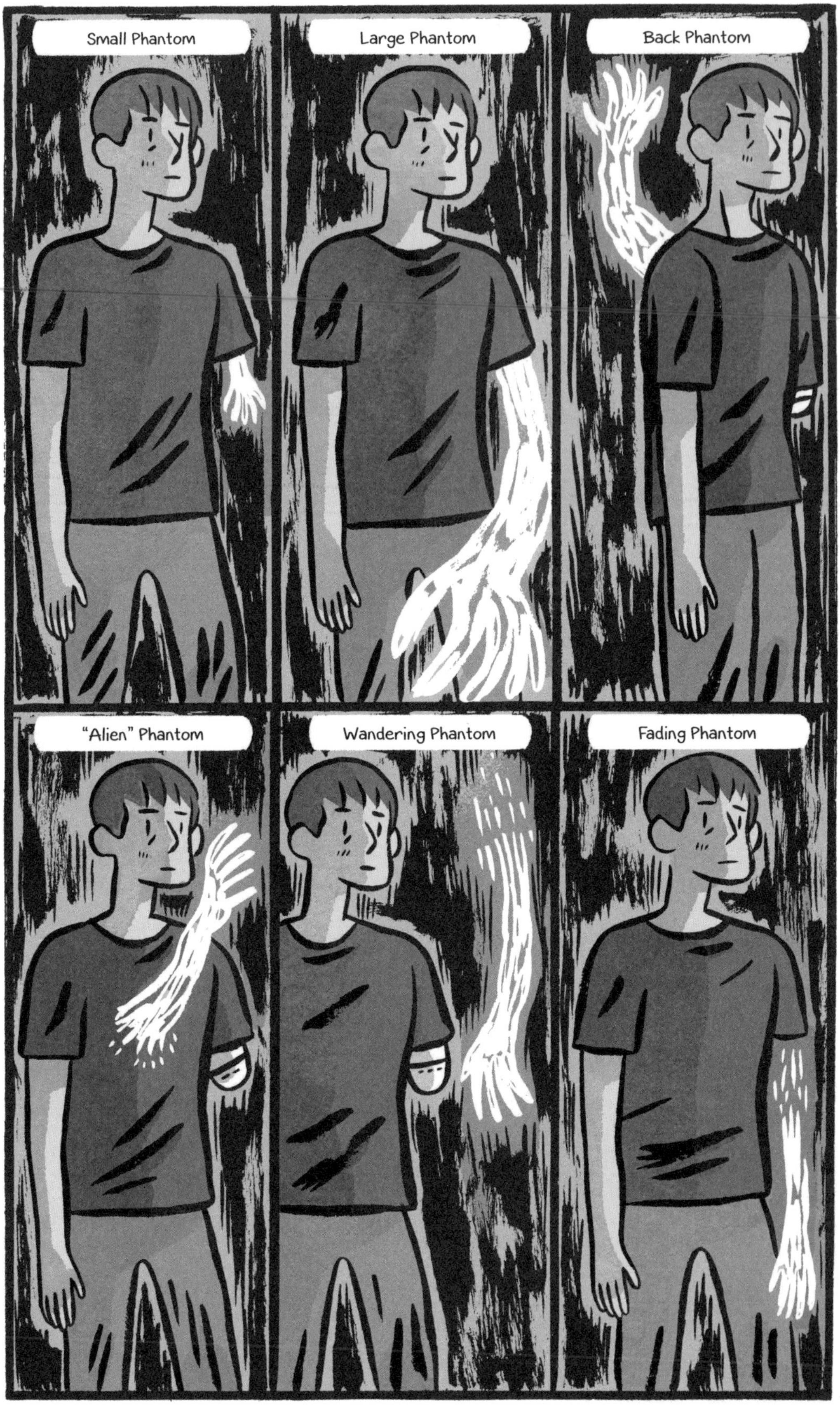
Small Phantom
Large Phantom
Back Phantom
"Alien" Phantom
Wandering Phantom
Fading Phantom

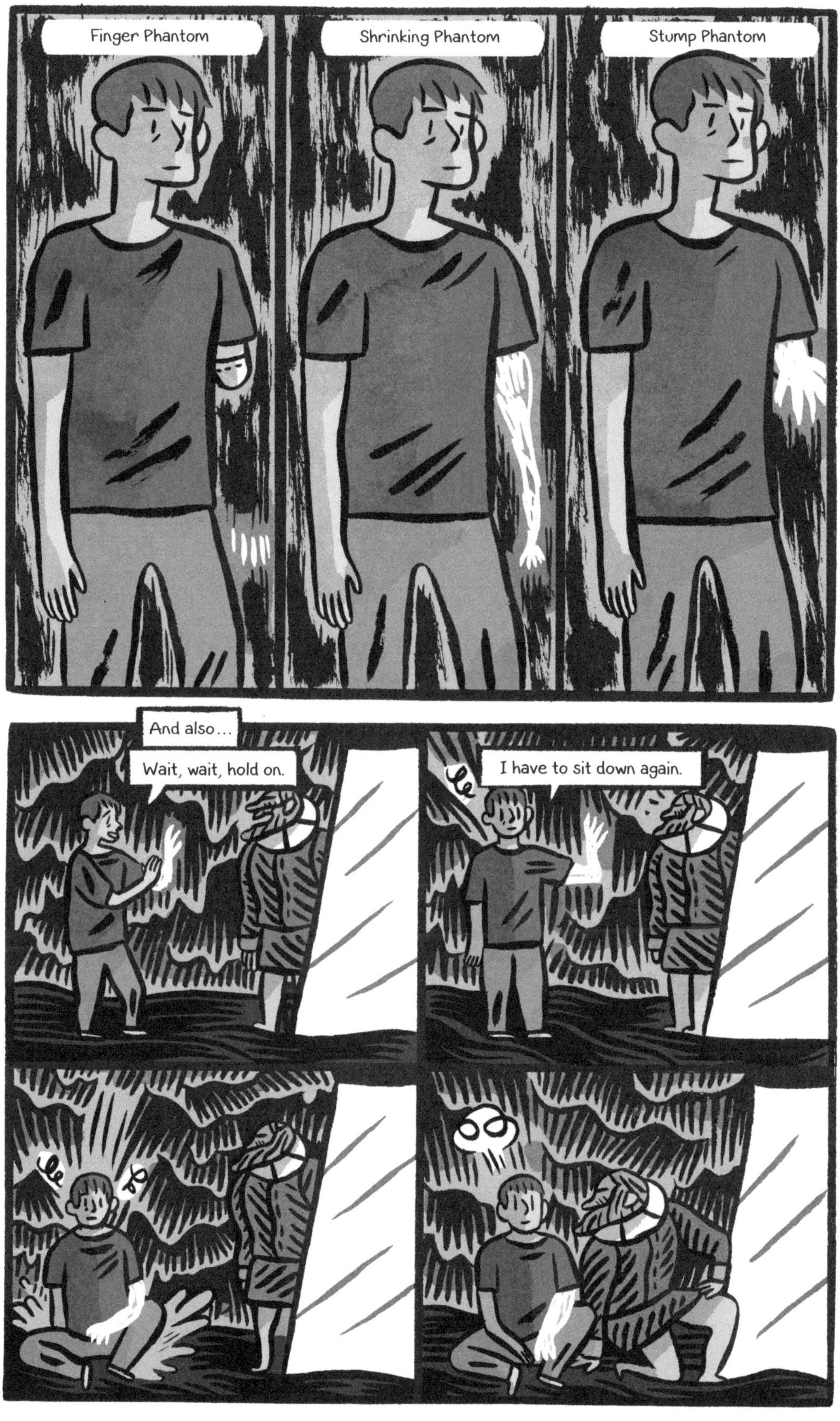
Finger Phantom
Shrinking Phantom
Stump Phantom
And also…
Wait, wait, hold on.
I have to sit down again.

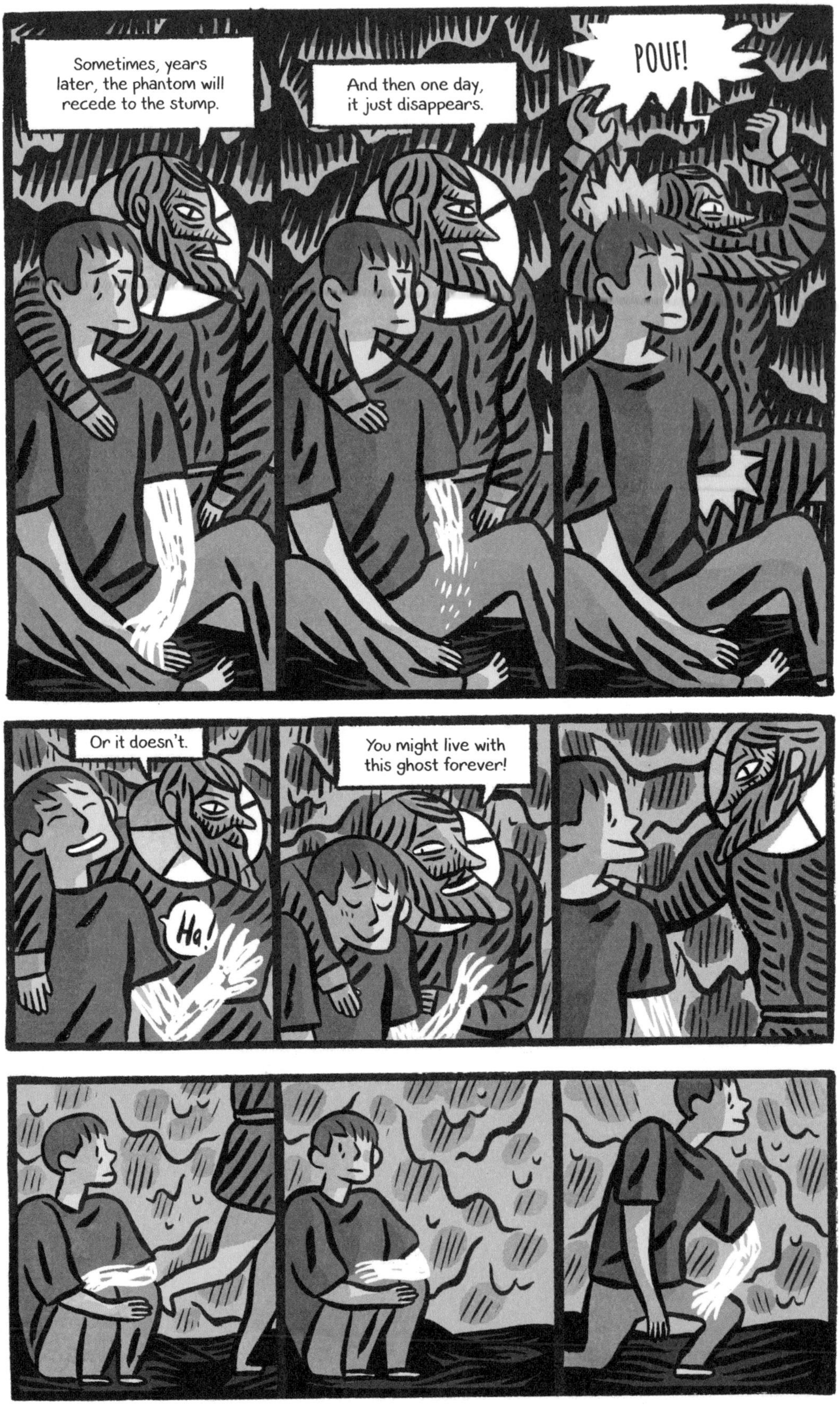
Sometimes, years later, the phantom will recede to the stump.
And then one day, it just disappears.
POUF!
Or it doesn't.
Ha!
You might live with this ghost forever!

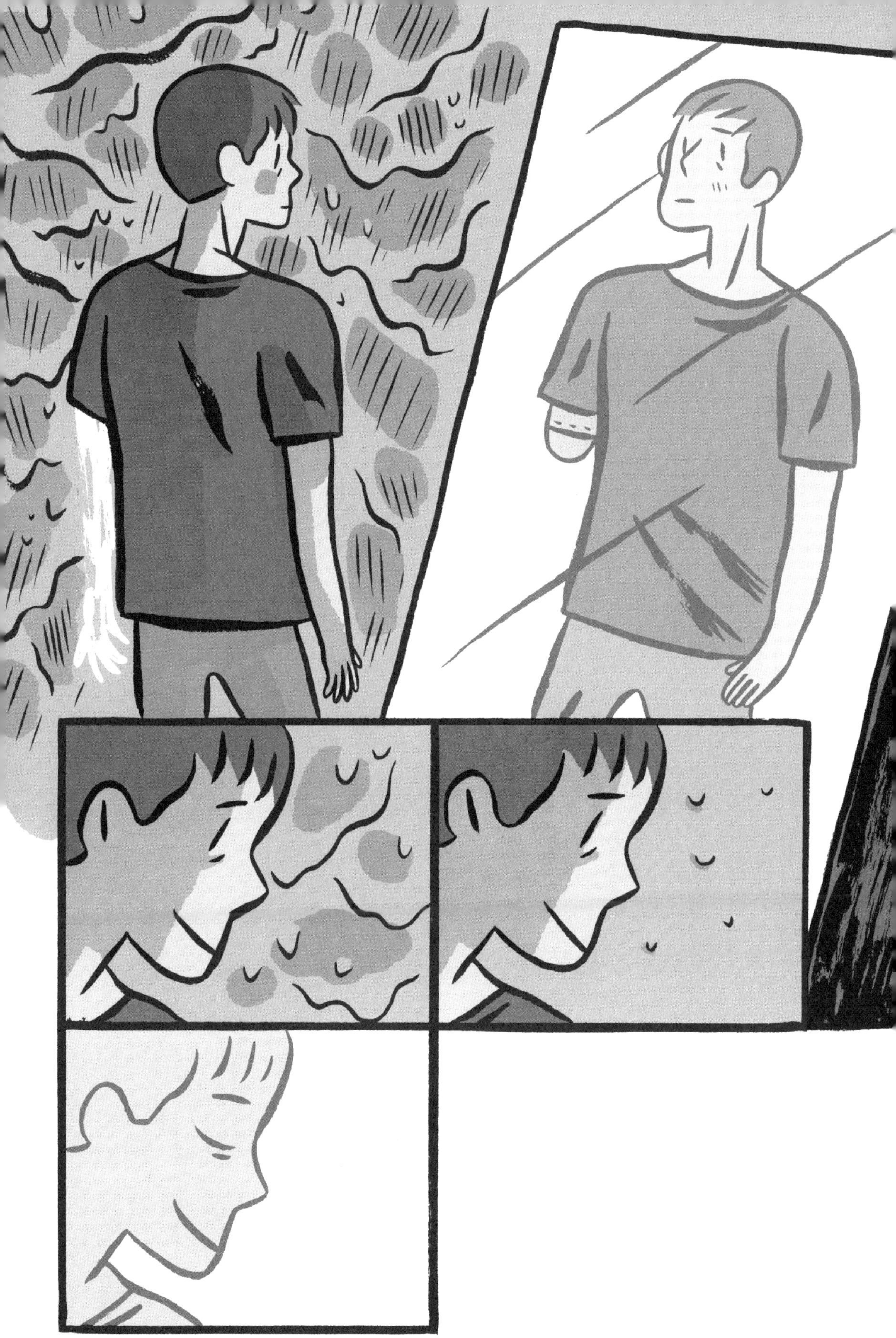

BRIEF INTERLUDE - THE BRAIN

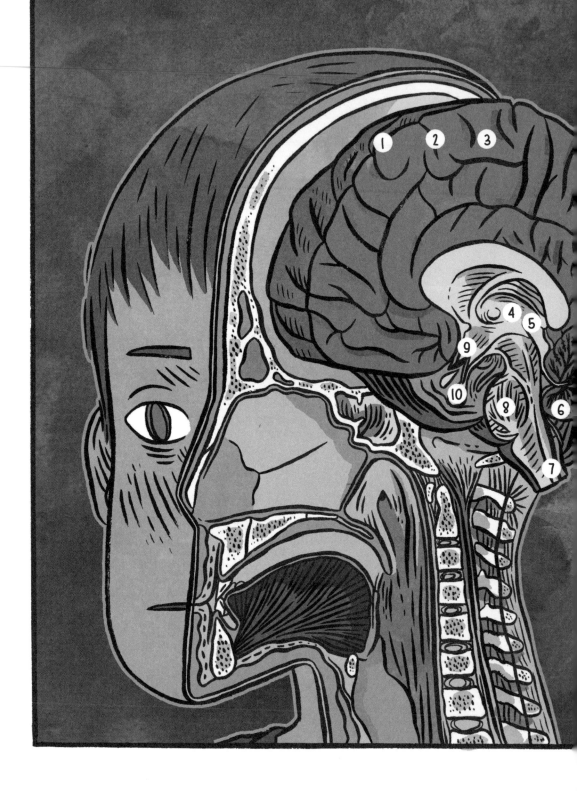

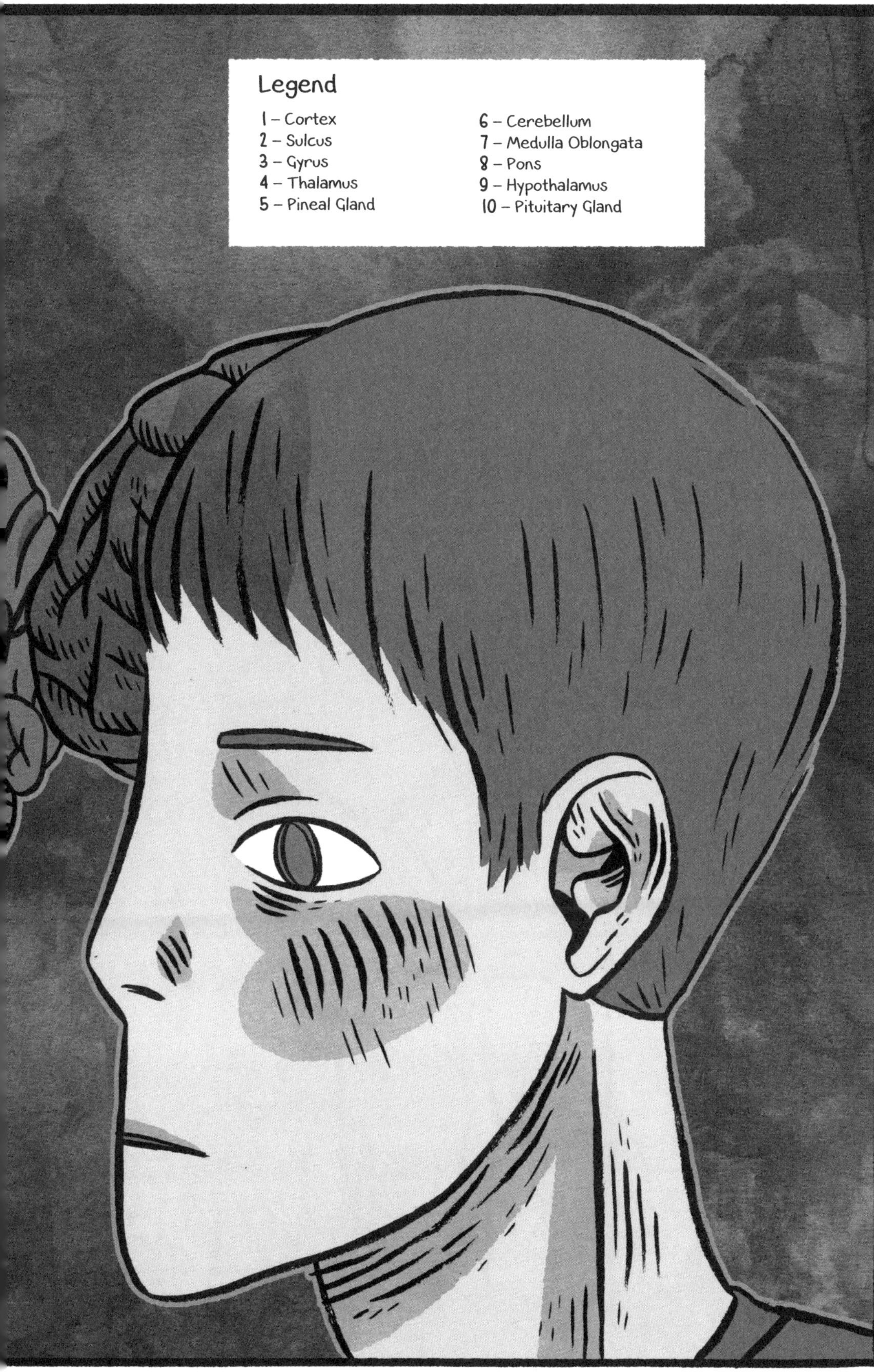
Legend
1 – Cortex
2 – Sulcus
3 – Gyrus
4 – Thalamus
5 – Pineal Gland
6 – Cerebellum
7 – Medulla Oblongata
8 – Pons
9 – Hypothalamus
10 – Pituitary Gland

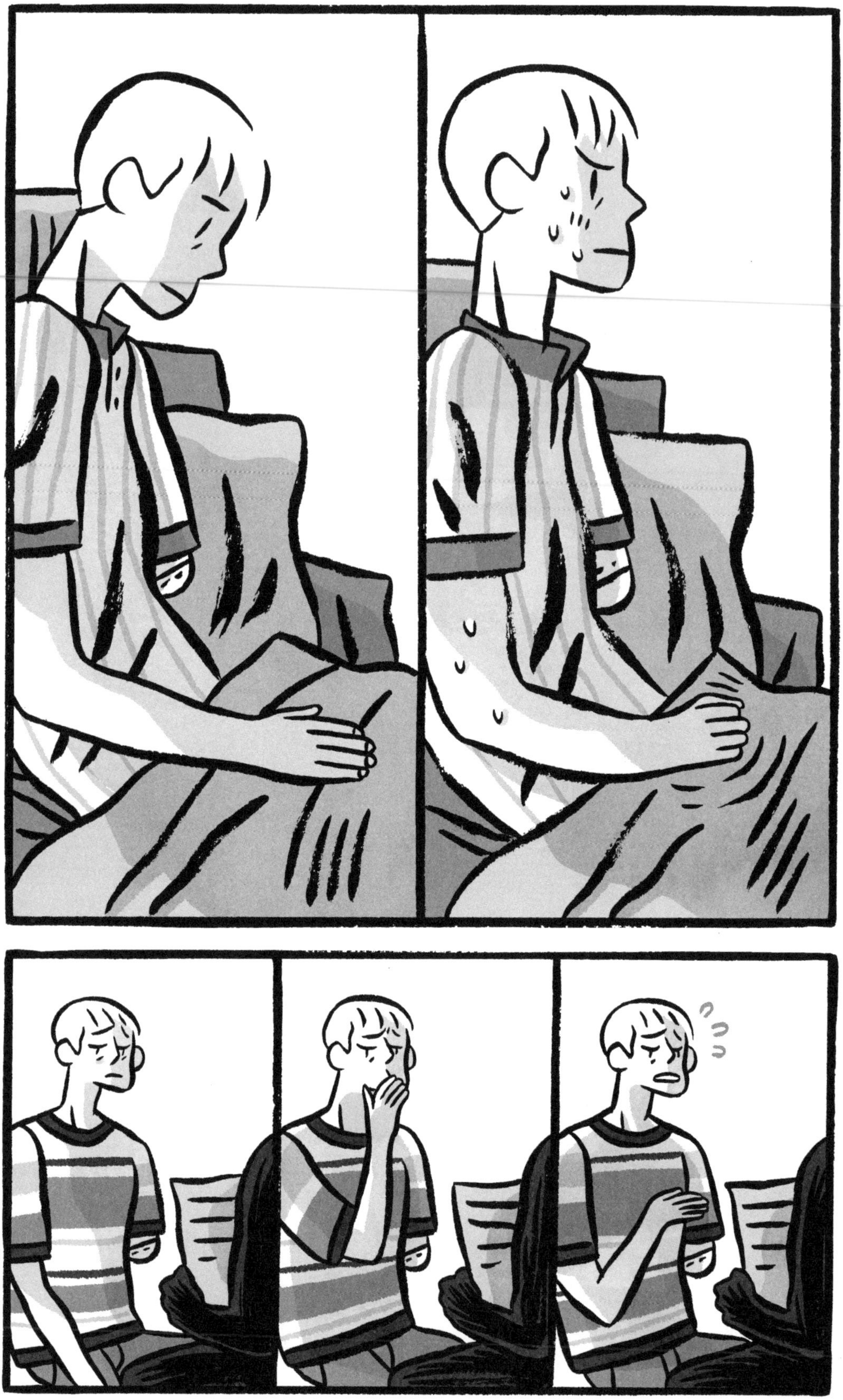

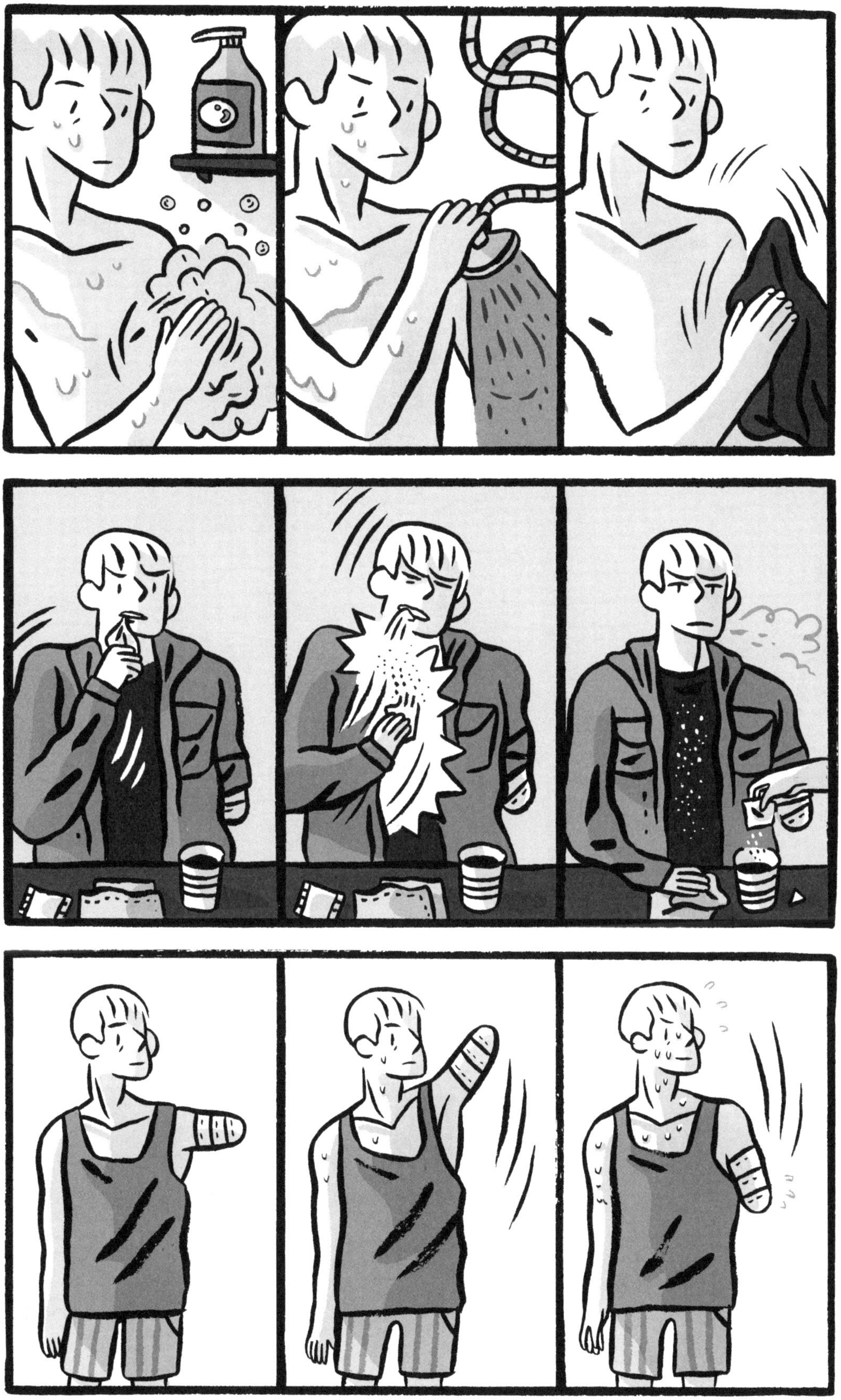

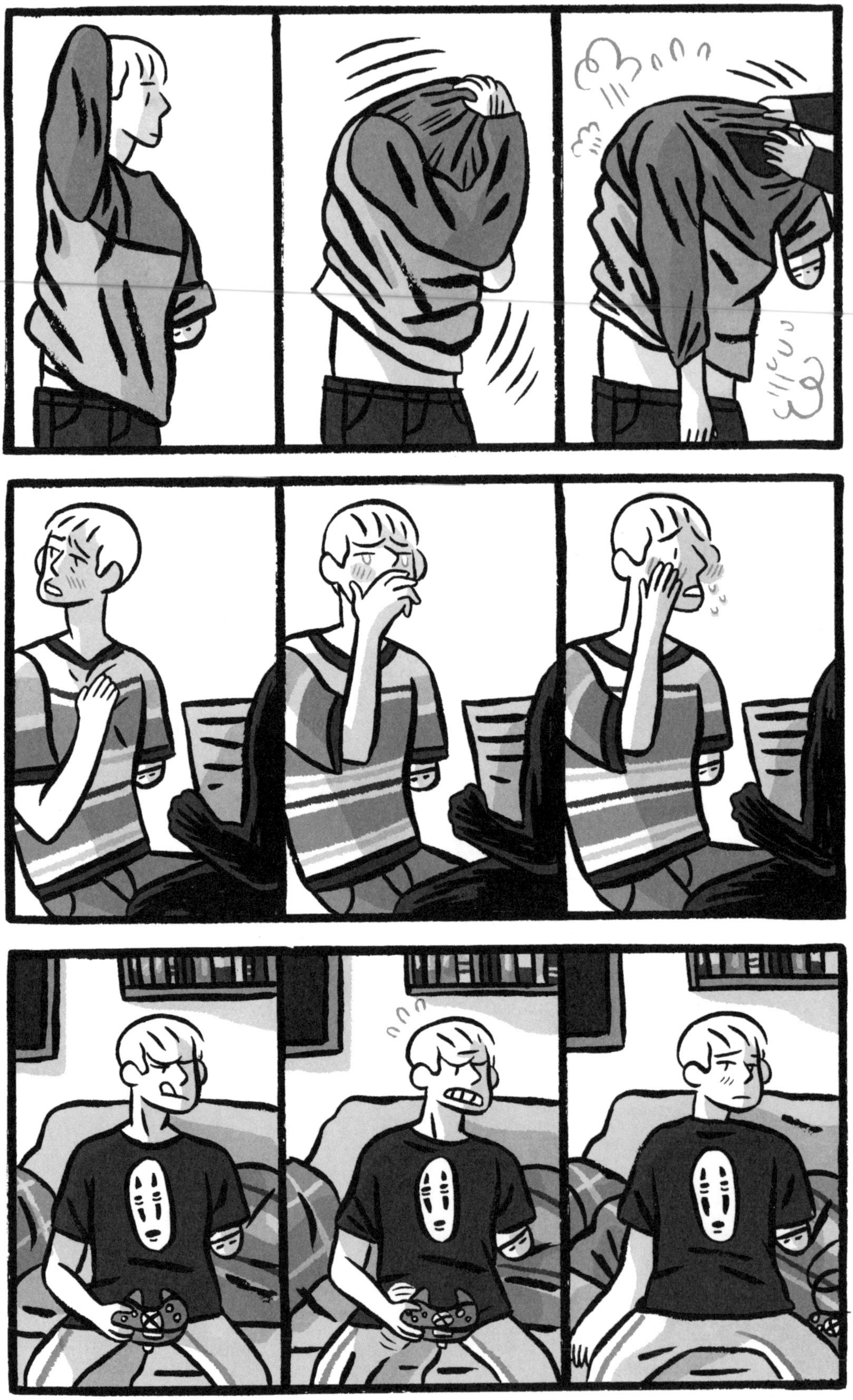

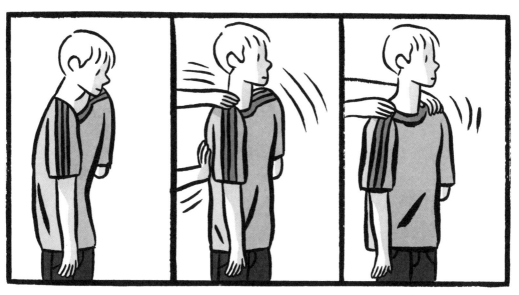

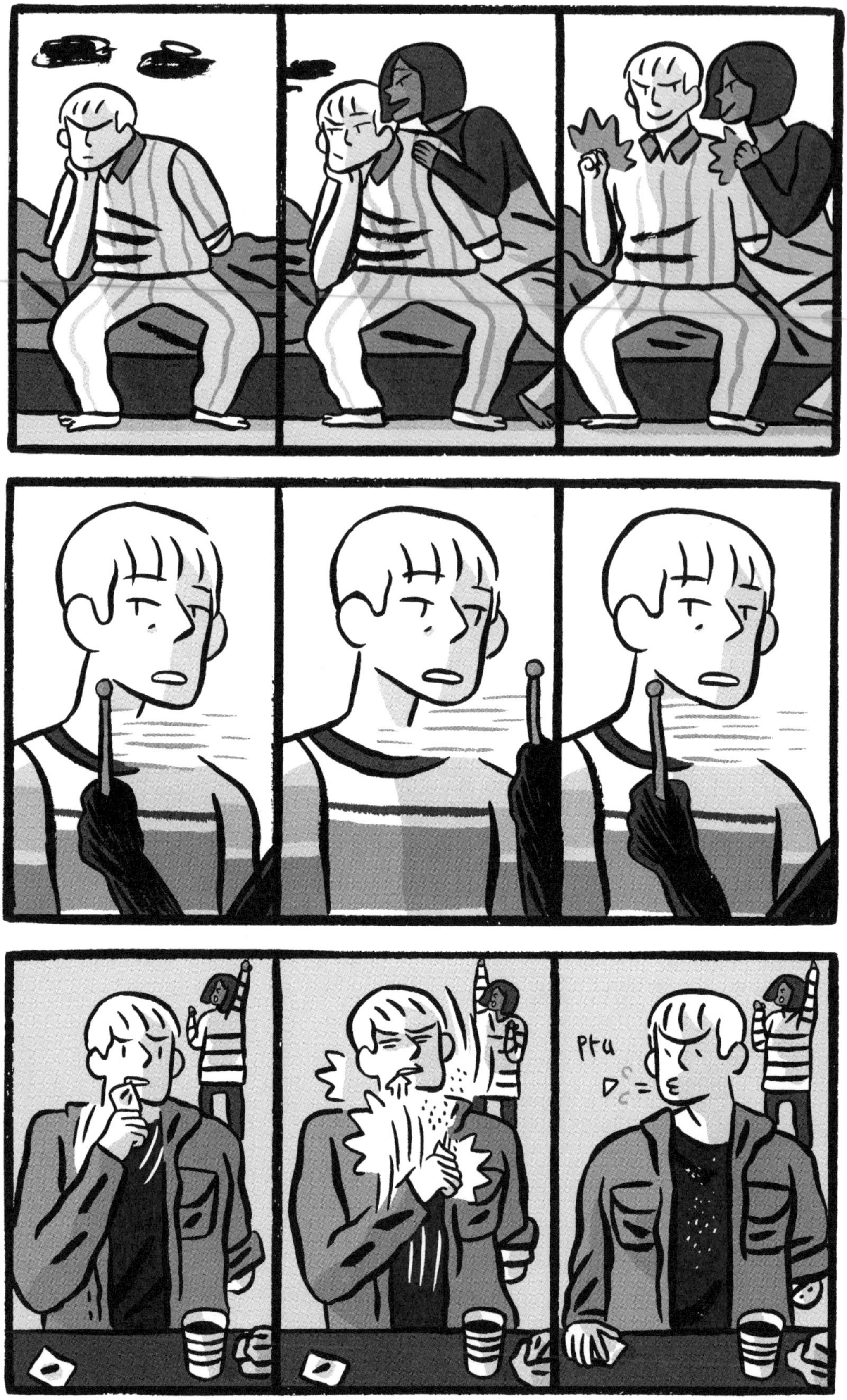
Ptu

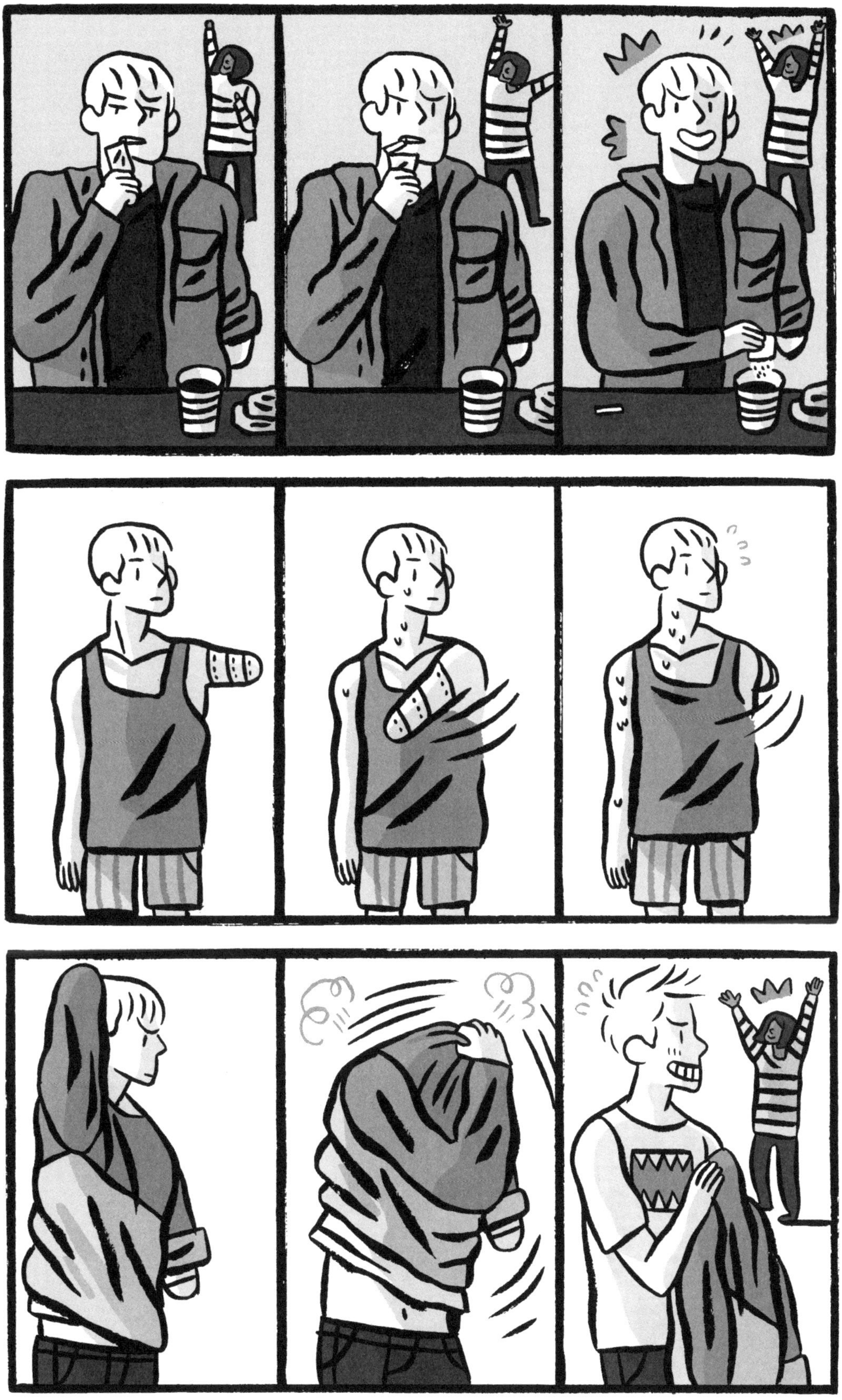

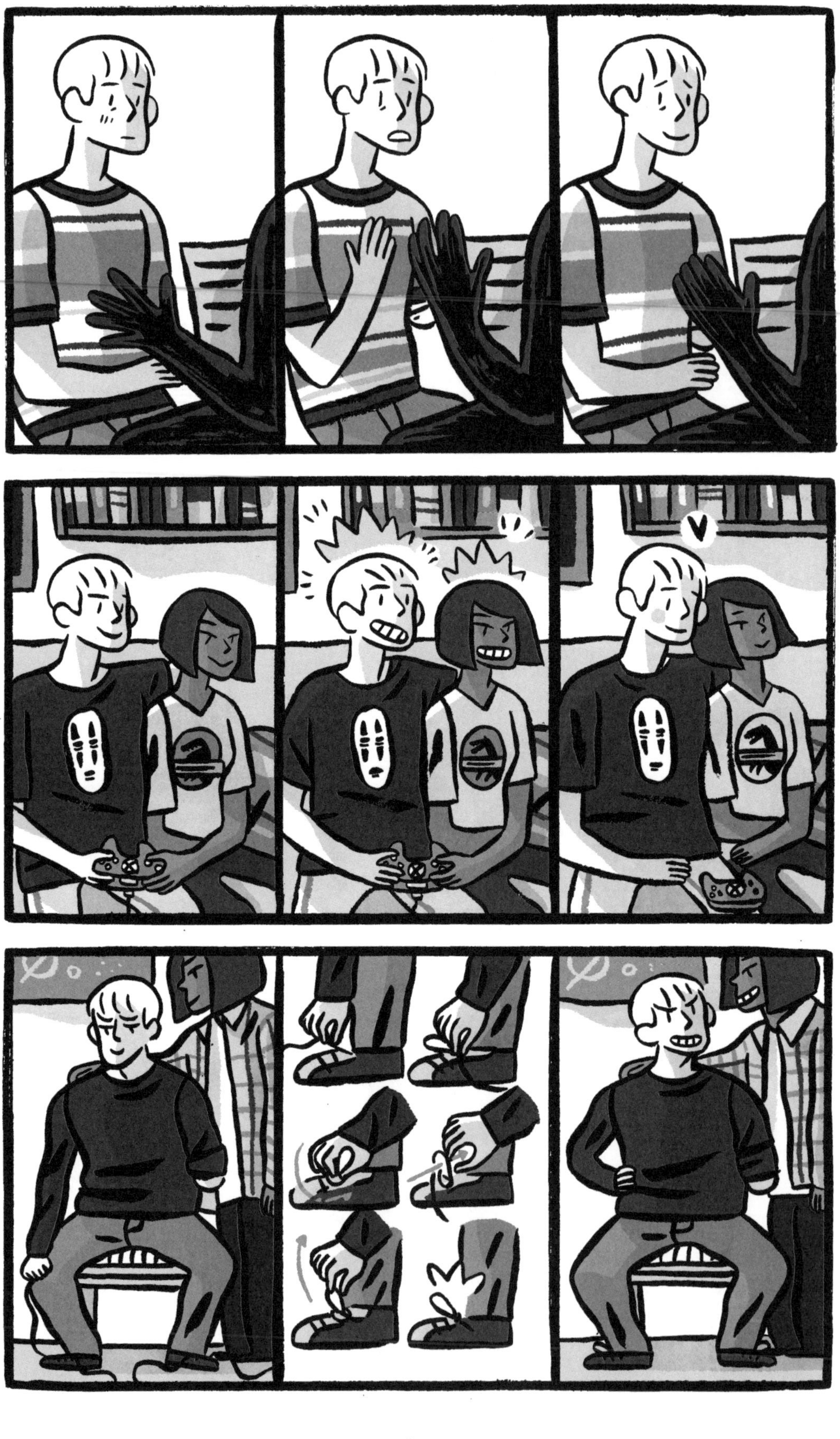

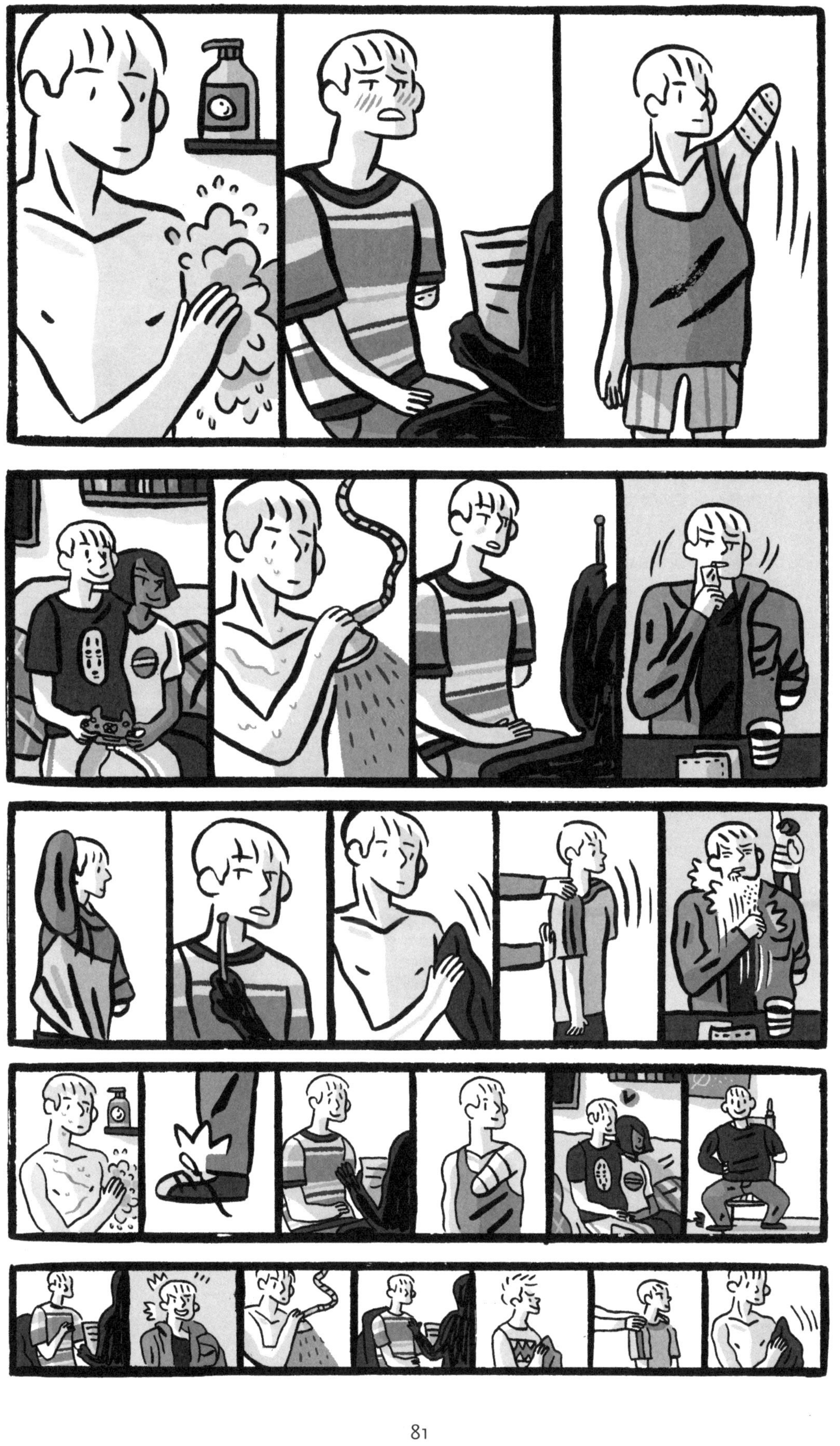

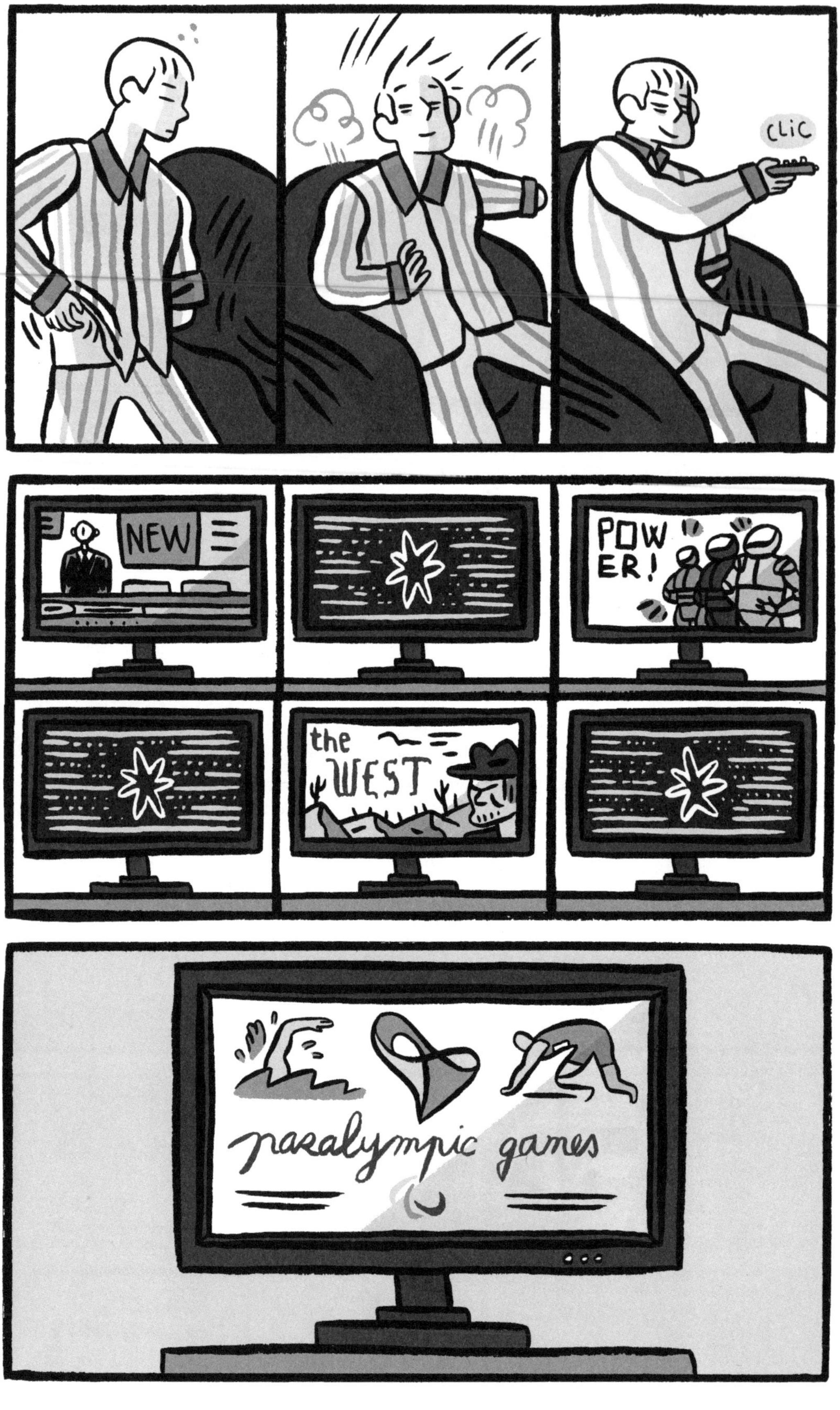
CLIC
NEW
POW ER!
the WEST
paralympic games

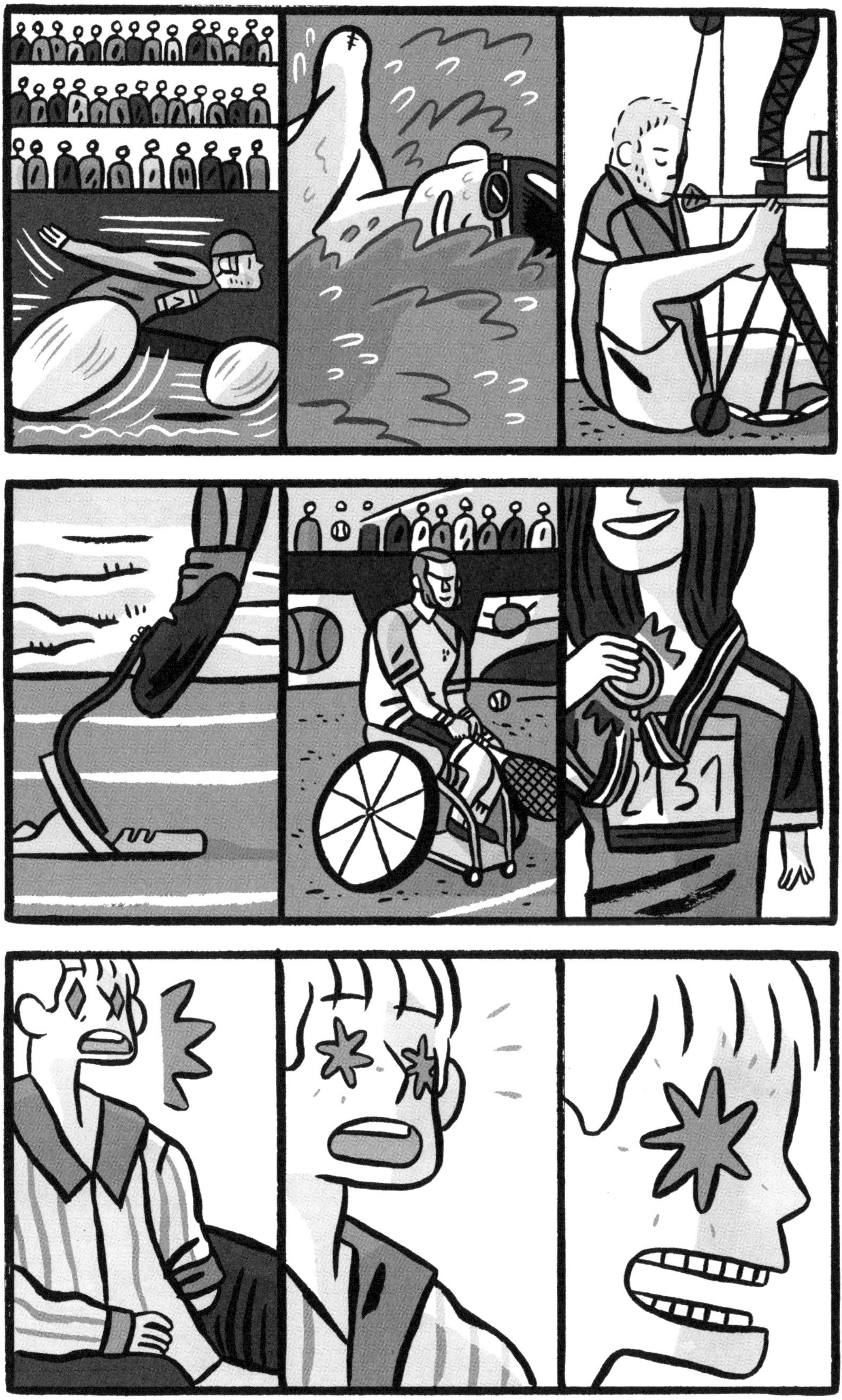

CHAPTER 3
Prostheses

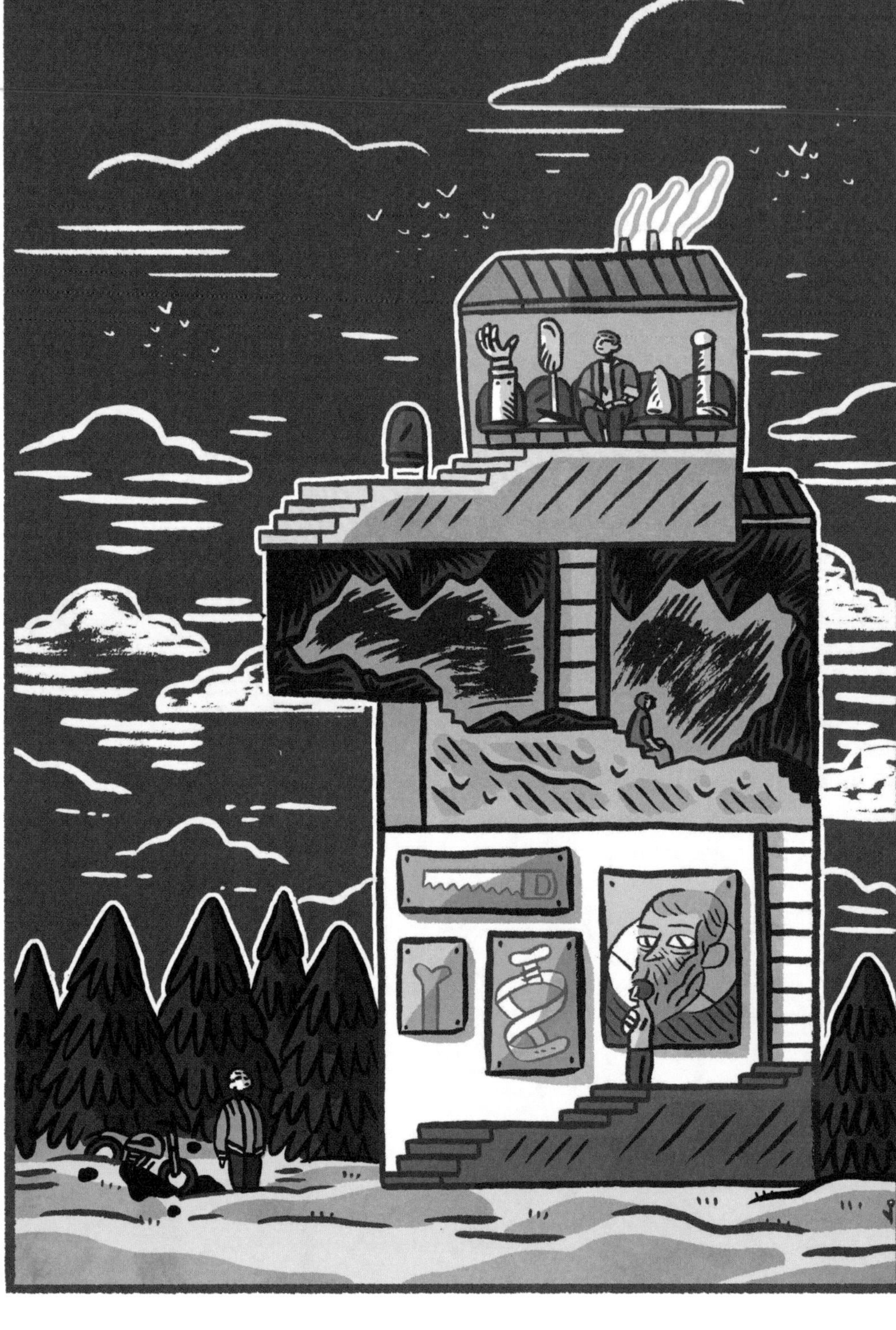

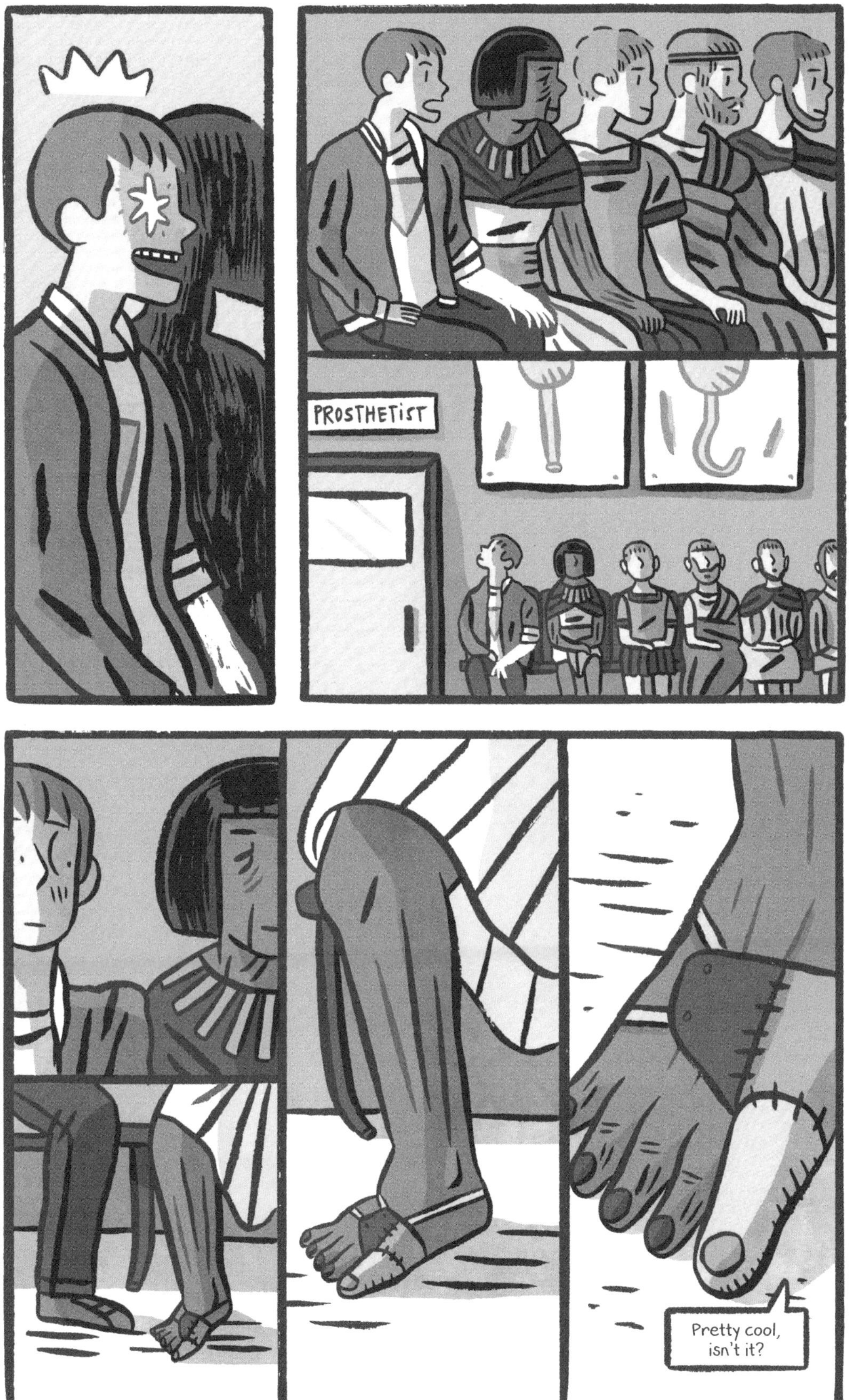
PROSTHETIST
Pretty cool, isn't it?

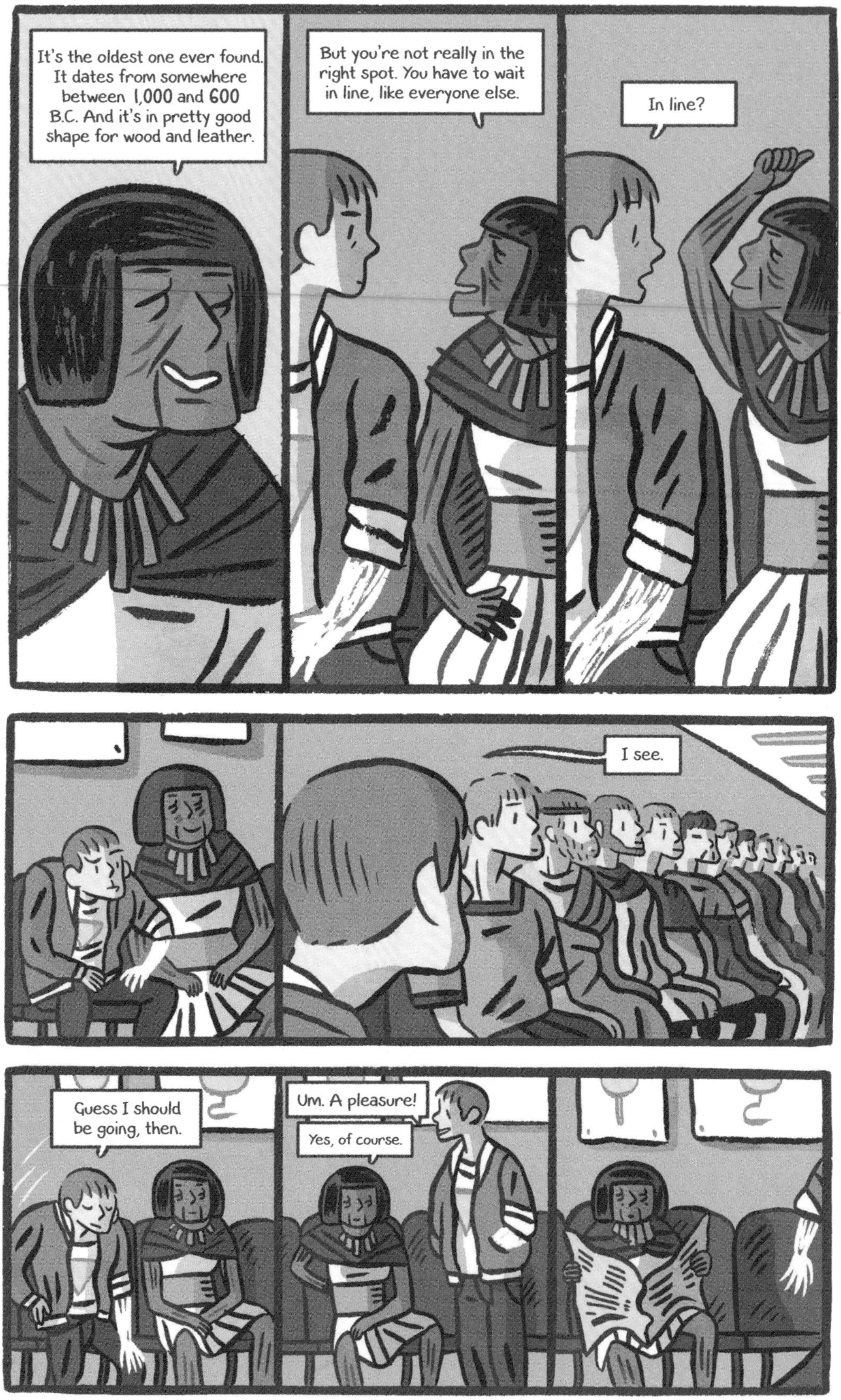
It's the oldest one ever found. It dates from somewhere between 1,000 and 600 B.C. And it's in pretty good shape for wood and leather.
But you're not really in the right spot. You have to wait in line, like everyone else.
In line?
I see.
Guess I should be going, then.
Um. A pleasure!
Yes, of course.

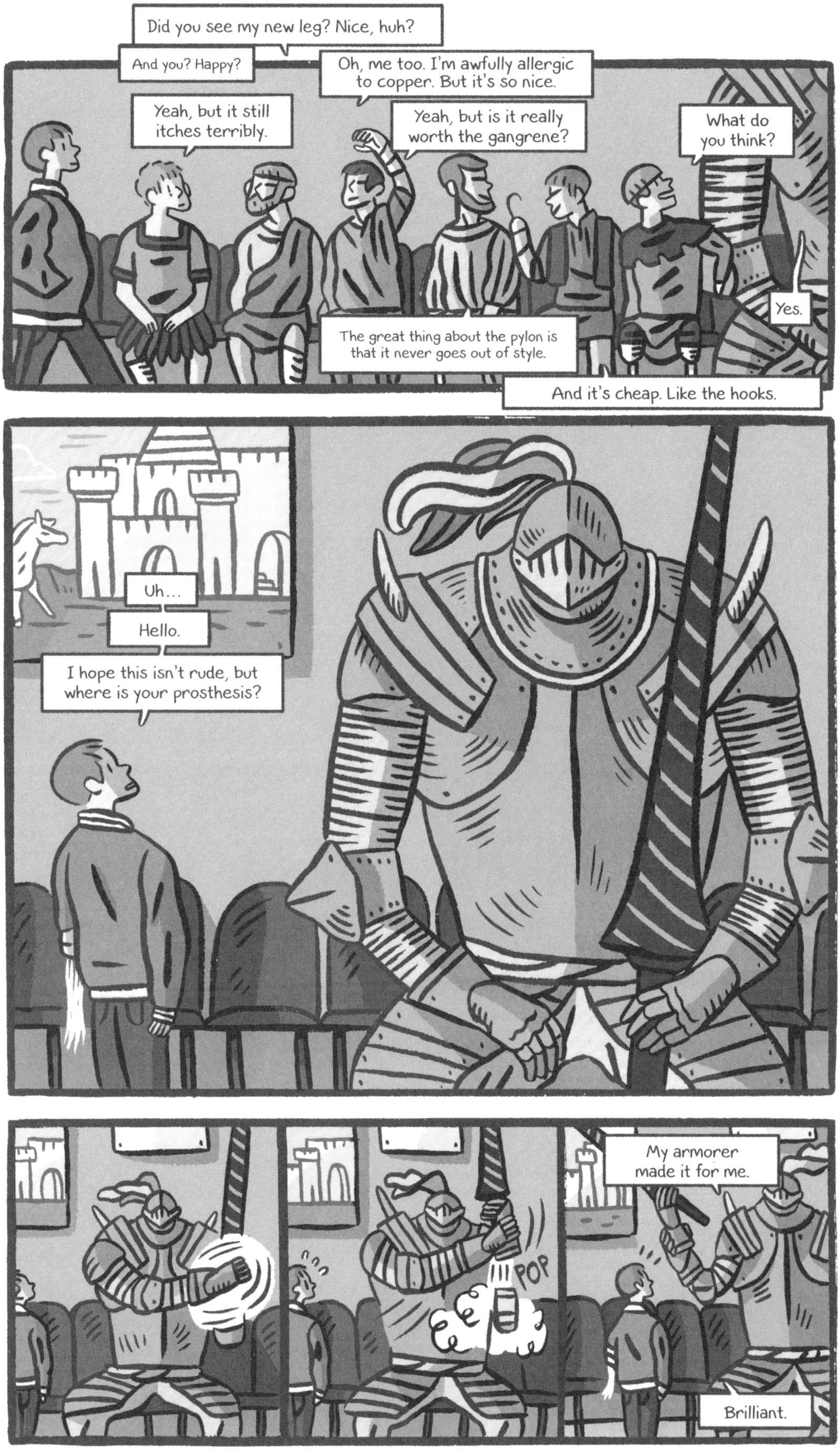

Did you see my new leg? Nice, huh?
And you? Happy?
Yeah, but it still itches terribly.
Oh, me too. I'm awfully allergic to copper. But it's so nice.
Yeah, but is it really worth the gangrene?
What do you think?
Yes.
The great thing about the pylon is that it never goes out of style.
And it's cheap. Like the hooks.
Uh...
Hello.
I hope this isn't rude, but where is your prosthesis?
POP
My armorer made it for me.
Brilliant.

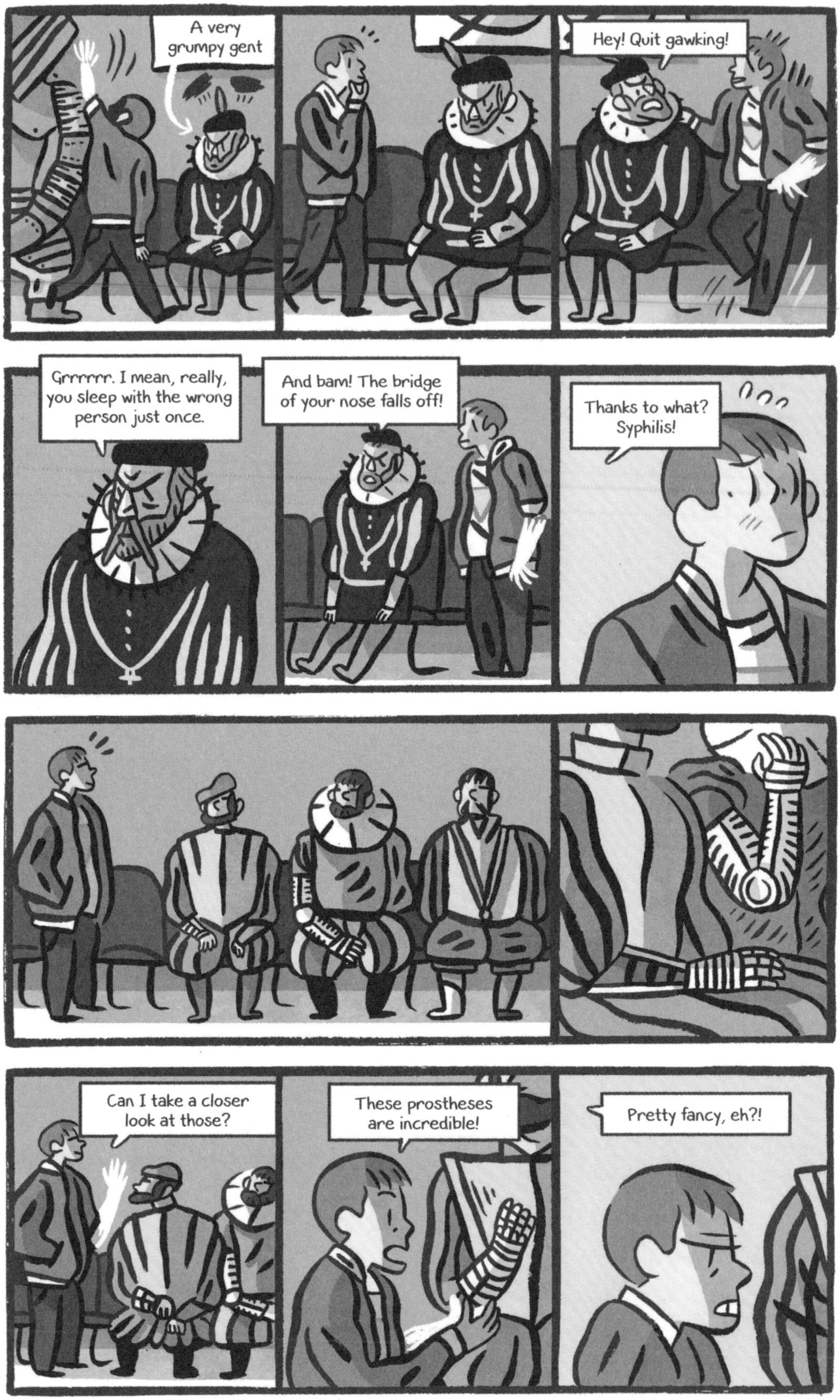
A very grumpy gent
Hey! Quit gawking!
Grrrrrr. I mean, really, you sleep with the wrong person just once.
And bam! The bridge of your nose falls off!
Thanks to what? Syphilis!
Can I take a closer look at those?
These prostheses are incredible!
Pretty fancy, eh?!

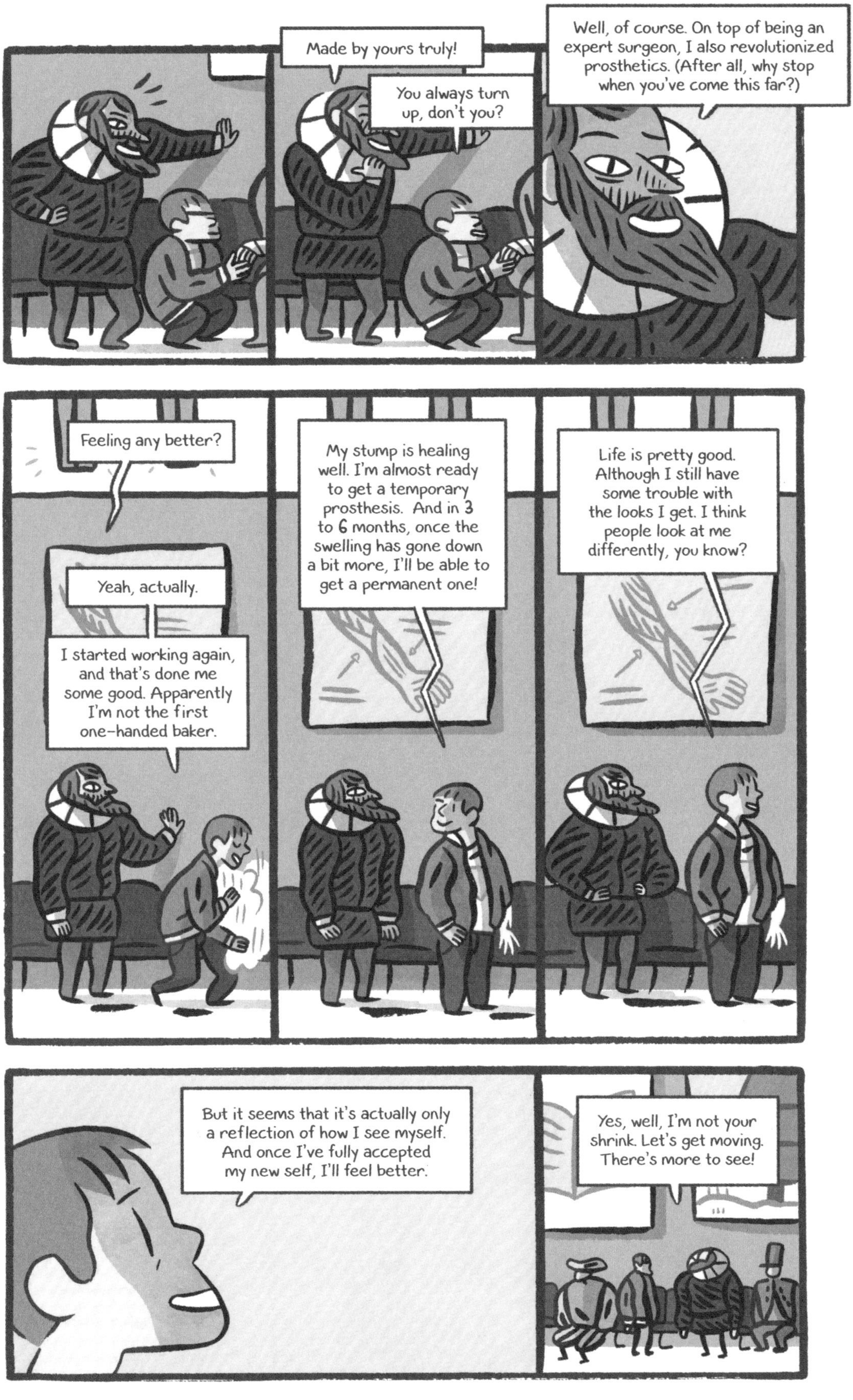
Made by yours truly!
You always turn up, don't you?
Well, of course. On top of being an expert surgeon, I also revolutionized prosthetics. (After all, why stop when you've come this far?)
Feeling any better?
Yeah, actually.
I started working again, and that's done me some good. Apparently I'm not the first one-handed baker.
My stump is healing well. I'm almost ready to get a temporary prosthesis. And in 3 to 6 months, once the swelling has gone down a bit more, I'll be able to get a permanent one!
Life is pretty good. Although I still have some trouble with the looks I get. I think people look at me differently, you know?
But it seems that it's actually only a reflection of how I see myself. And once I've fully accepted my new self, I'll feel better.
Yes, well, I'm not your shrink. Let's get moving. There's more to see!

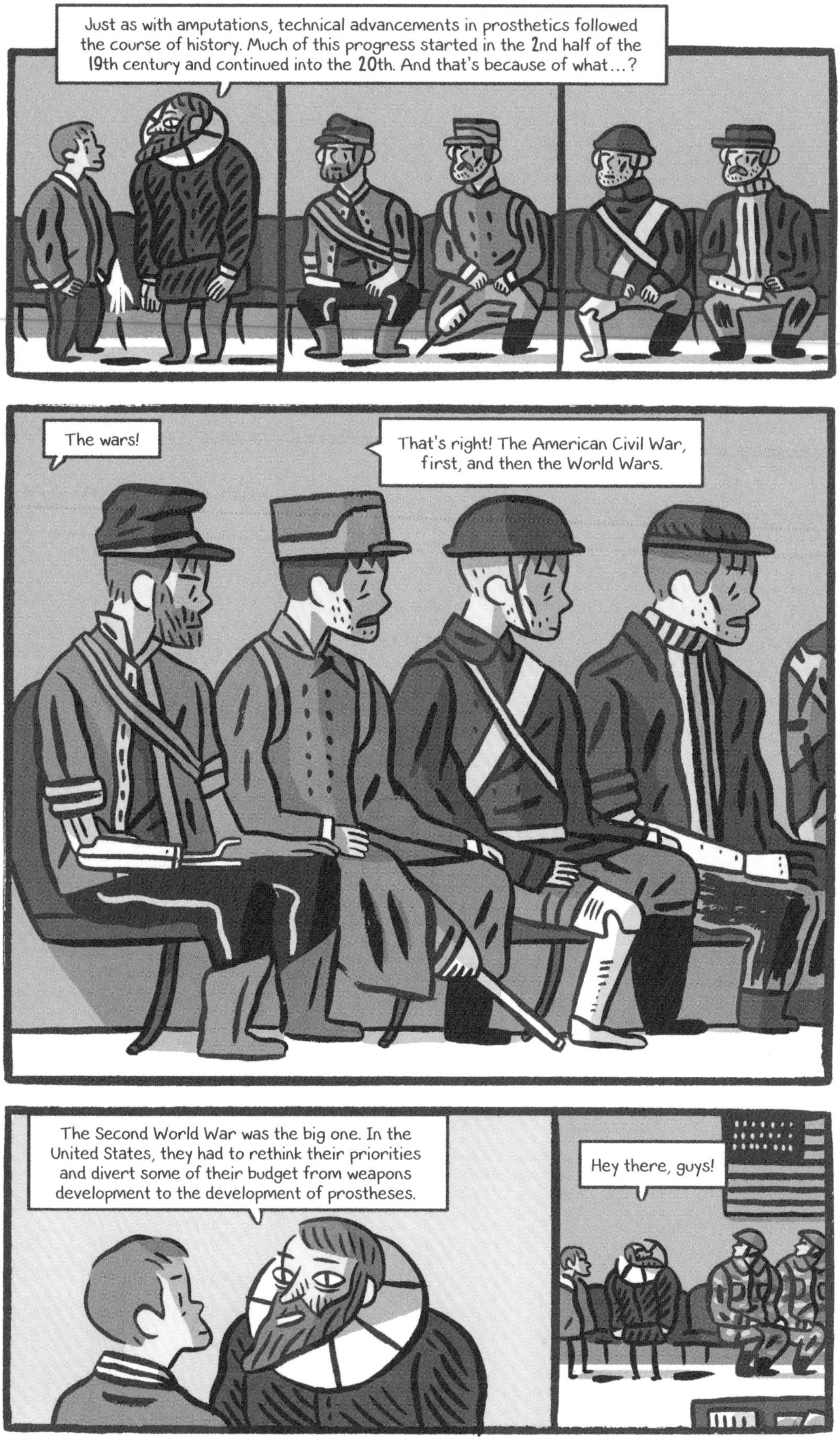
Just as with amputations, technical advancements in prosthetics followed the course of history. Much of this progress started in the 2nd half of the 19th century and continued into the 20th. And that's because of what...?
The wars!
That's right! The American Civil War, first, and then the World Wars.
The Second World War was the big one. In the United States, they had to rethink their priorities and divert some of their budget from weapons development to the development of prostheses.
Hey there, guys!

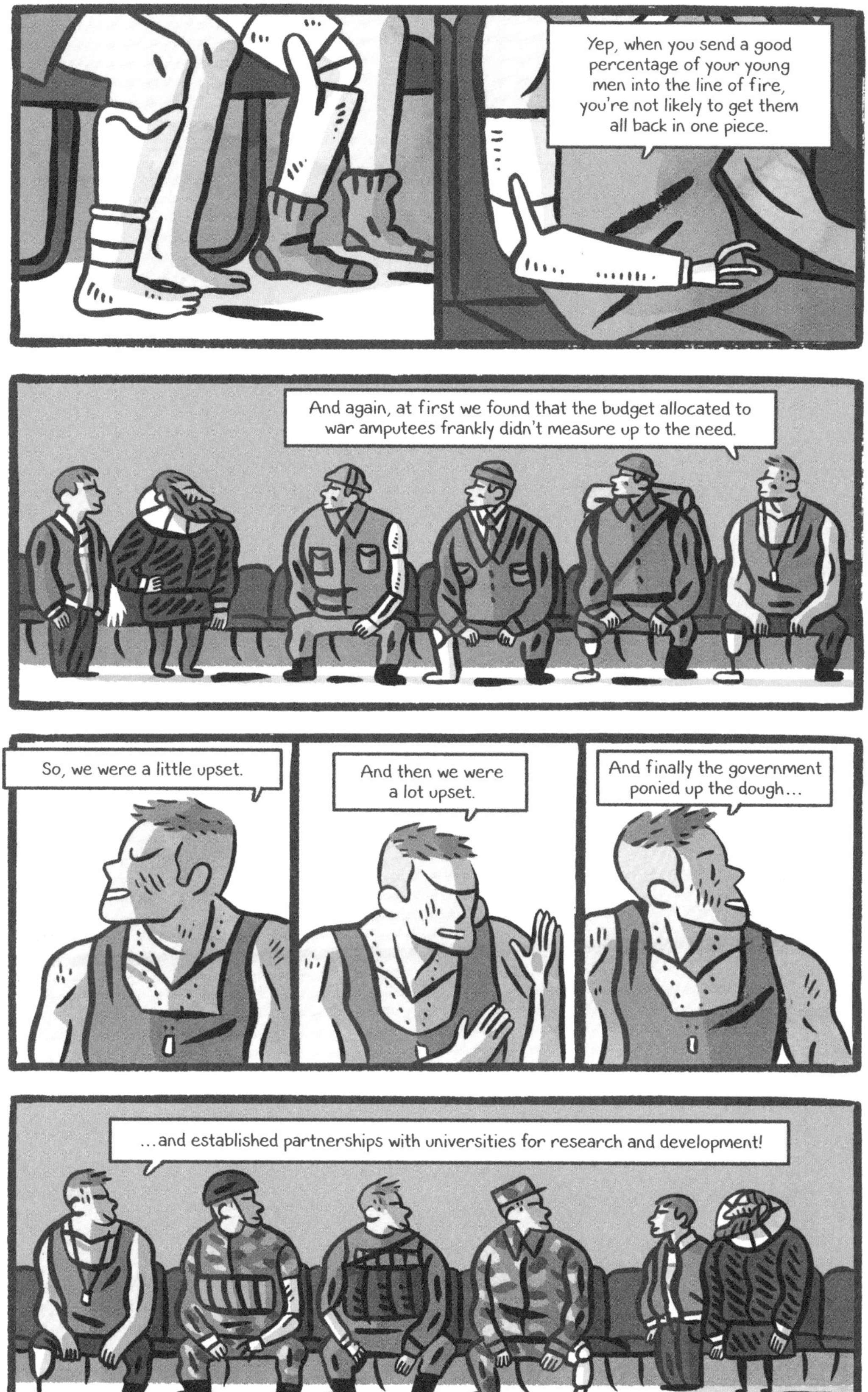
Yep, when you send a good percentage of your young men into the line of fire, you're not likely to get them all back in one piece.
And again, at first we found that the budget allocated to war amputees frankly didn't measure up to the need.
So, we were a little upset.
And then we were a lot upset.
And finally the government ponied up the dough…
…and established partnerships with universities for research and development!

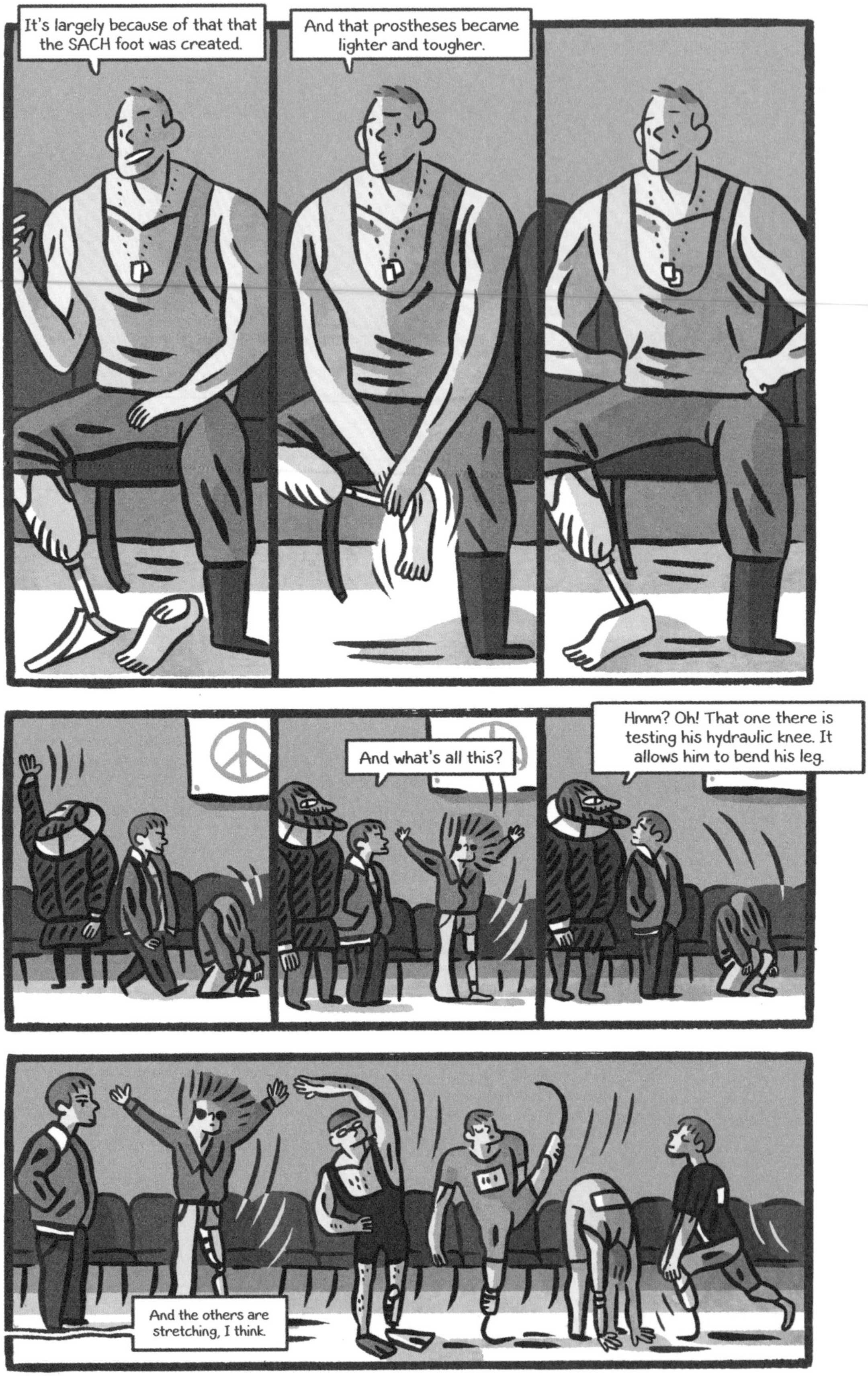
It's largely because of that that the SACH foot was created.
And that prostheses became lighter and tougher.
And what's all this?
Hmm? Oh! That one there is testing his hydraulic knee. It allows him to bend his leg.
And the others are stretching, I think.

Today, we have prostheses that have been developed for specific sports—swimming, running, cycling…As a result, both the practice of sport for athletes with disabilities and their comfort while doing so have greatly improved.

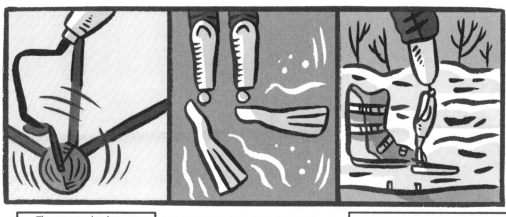

The progress has been so great that some athletes have participated in the Olympics alongside more able-bodied athletes.

Like the infamous Oscar Pistorius, in 2012.

All that said, we're still far from a future where technology will be able to level the playing field between athletes with artificial limbs and those without.

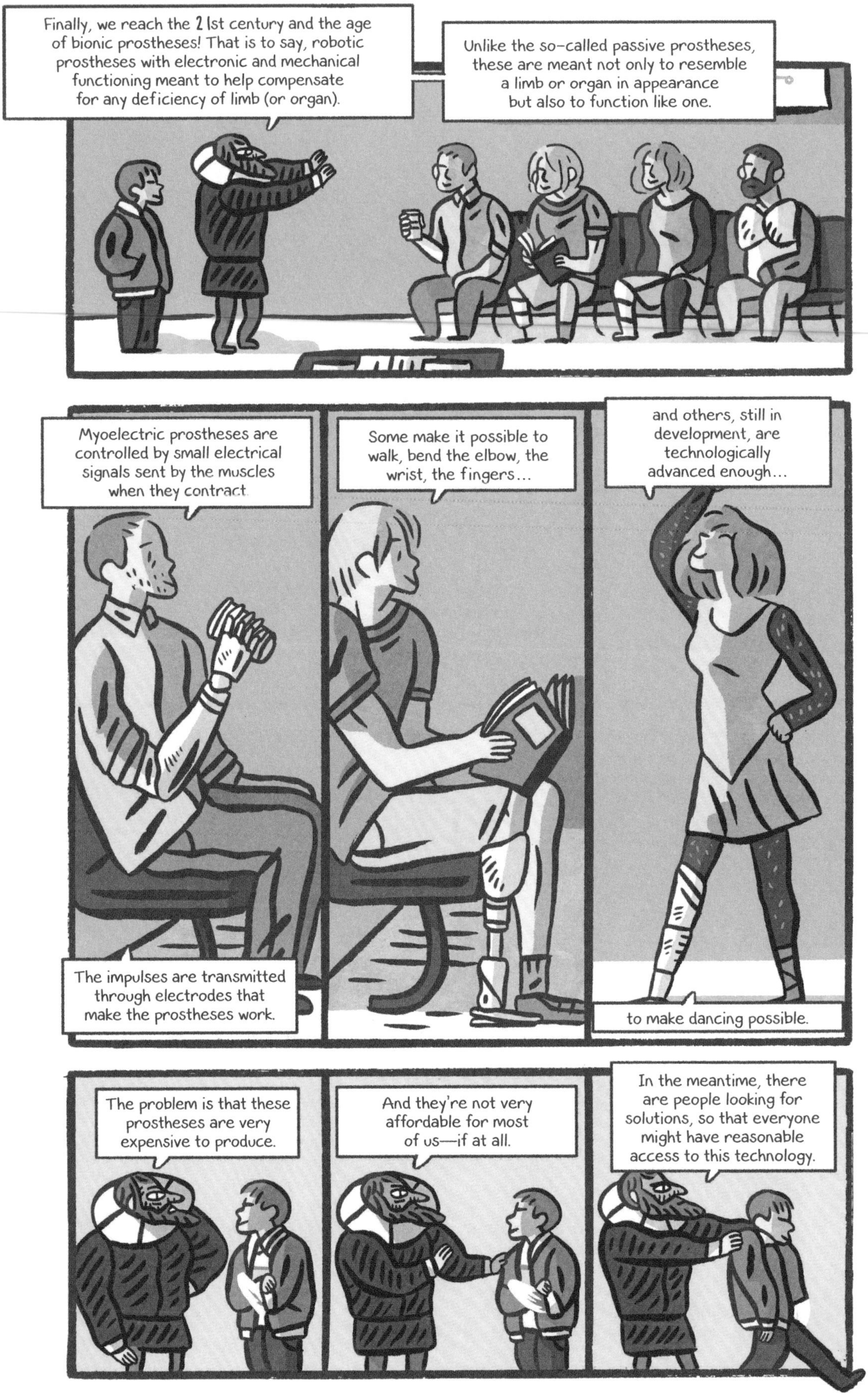

Finally, we reach the 21st century and the age of bionic prostheses! That is to say, robotic prostheses with electronic and mechanical functioning meant to help compensate for any deficiency of limb (or organ).
Unlike the so-called passive prostheses, these are meant not only to resemble a limb or organ in appearance but also to function like one.
Myoelectric prostheses are controlled by small electrical signals sent by the muscles when they contract.
The impulses are transmitted through electrodes that make the prostheses work.
Some make it possible to walk, bend the elbow, the wrist, the fingers…
and others, still in development, are technologically advanced enough…
to make dancing possible.
The problem is that these prostheses are very expensive to produce.
And they're not very affordable for most of us—if at all.
In the meantime, there are people looking for solutions, so that everyone might have reasonable access to this technology.

Some have built "fab labs" (like the ones put in place by My Human Kit, founded by Nicolas Huchet) and succeeded in designing and producing bionic hand prostheses, manufactured almost entirely with 3D printers, for only $230!

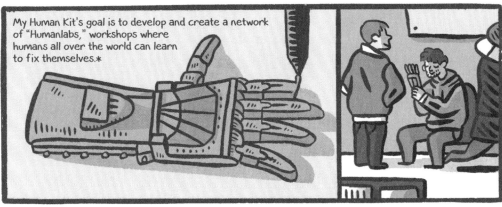

My Human Kit's goal is to develop and create a network of "Humanlabs," workshops where humans all over the world can learn to fix themselves.*

This is also true of organizations like Handicap International, which works to improve the quality of life for people disabled as a result of war and to help them function fully as members of society.

We've covered a lot, huh?!

* Blueprints and code are freely downloadable on their site (myhumankit.org)!

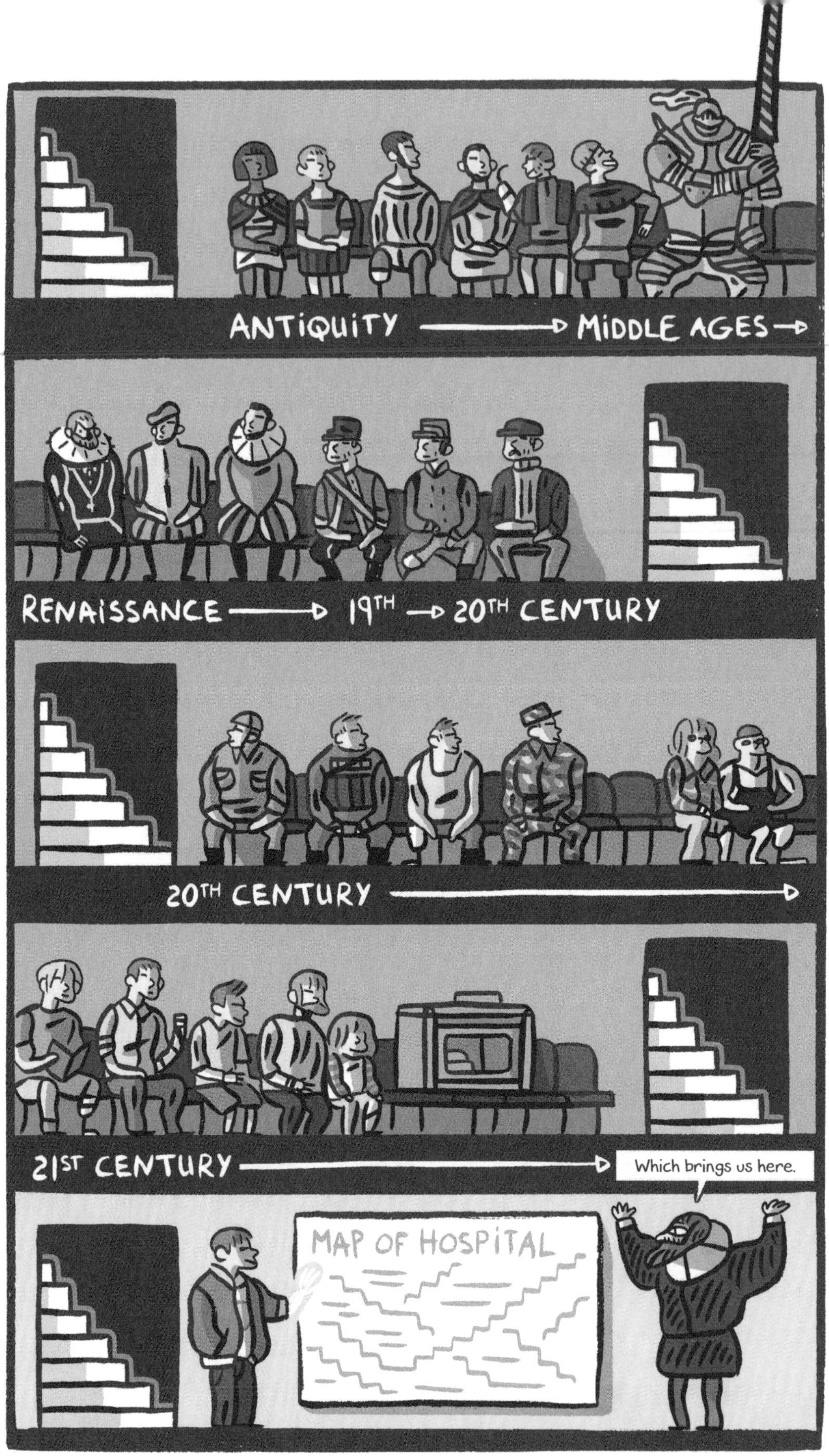
ANTIQUITY → MIDDLE AGES →
RENAISSANCE → 19TH → 20TH CENTURY
20TH CENTURY →
21ST CENTURY →
Which brings us here.
MAP OF HOSPITAL

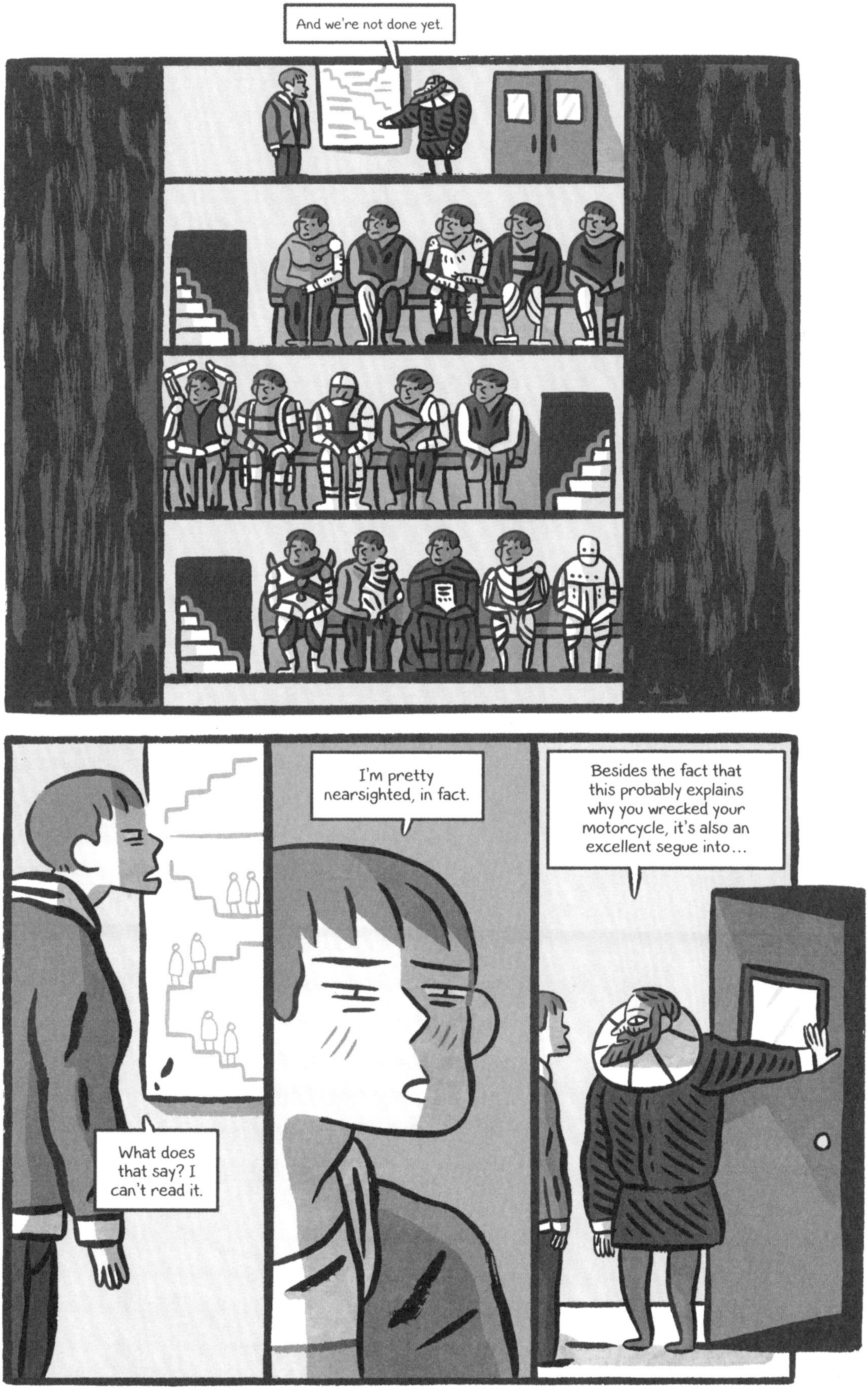
And we're not done yet.
What does that say? I can't read it.
I'm pretty nearsighted, in fact.
Besides the fact that this probably explains why you wrecked your motorcycle, it's also an excellent segue into…

A Partial List of Prostheses

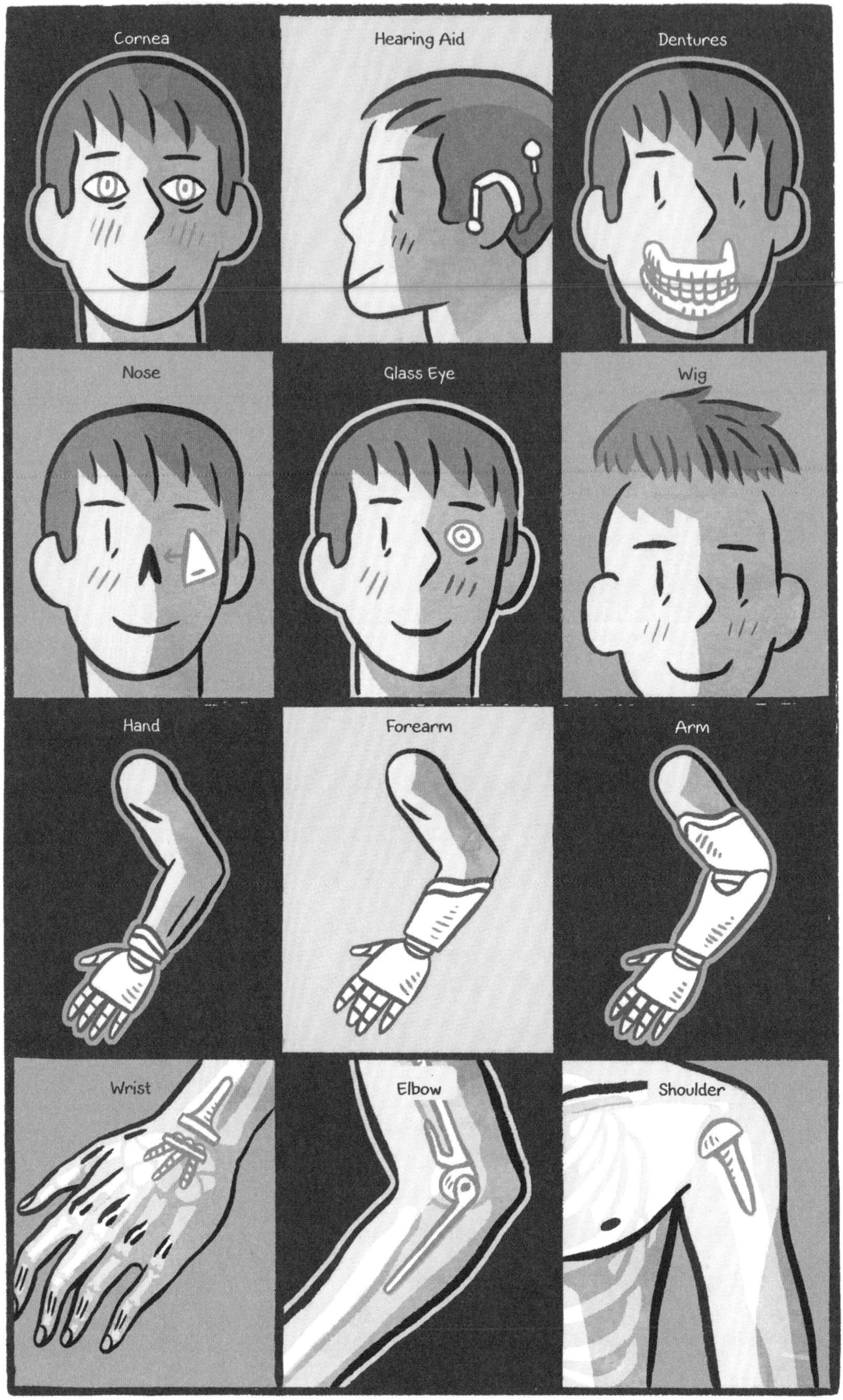
Cornea
Hearing Aid
Dentures
Nose
Glass Eye
Wig
Hand
Forearm
Arm
Wrist
Elbow
Shoulder

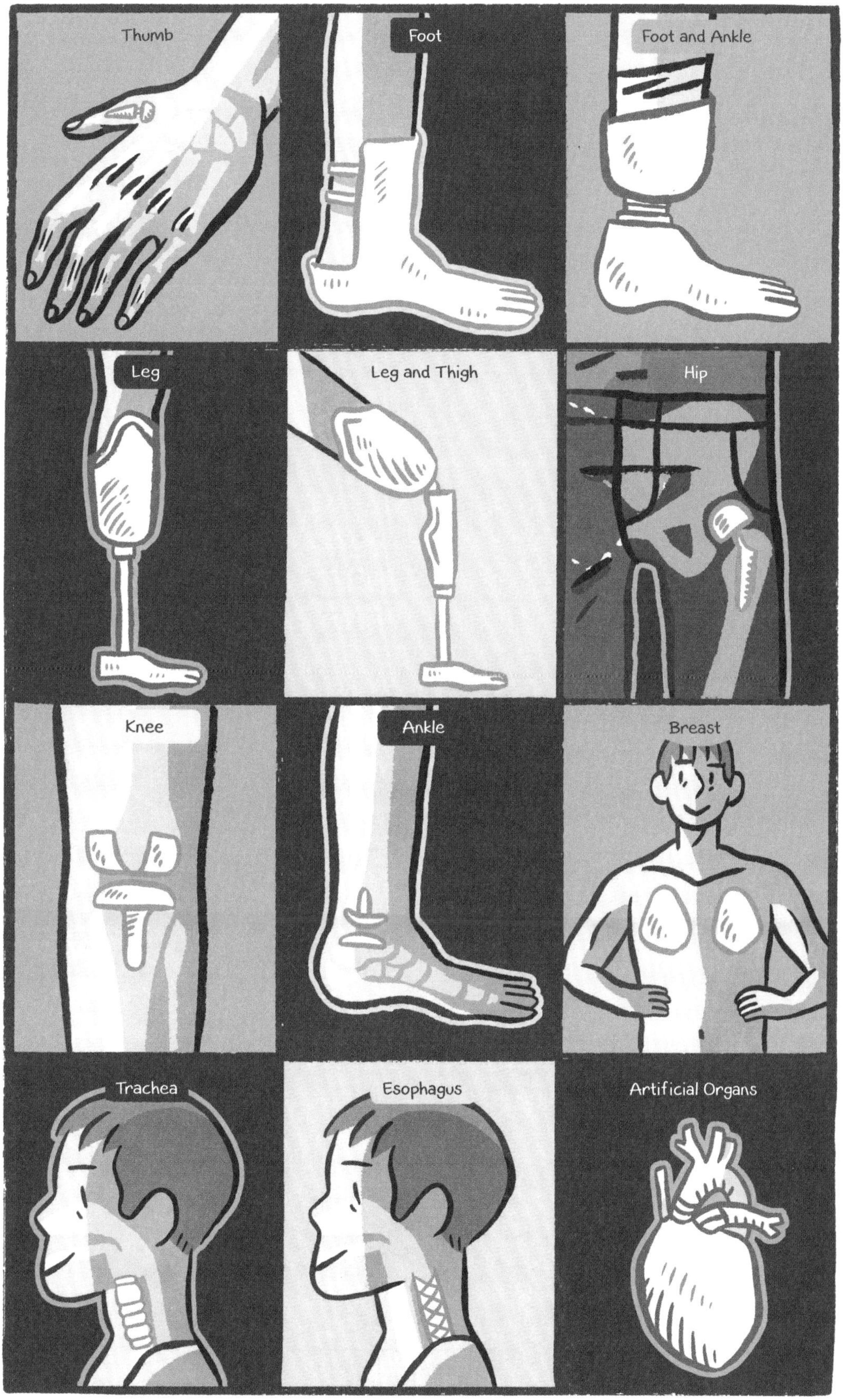
Thumb
Foot
Foot and Ankle
Leg
Leg and Thigh
Hip
Knee
Ankle
Breast
Trachea
Esophagus
Artificial Organs

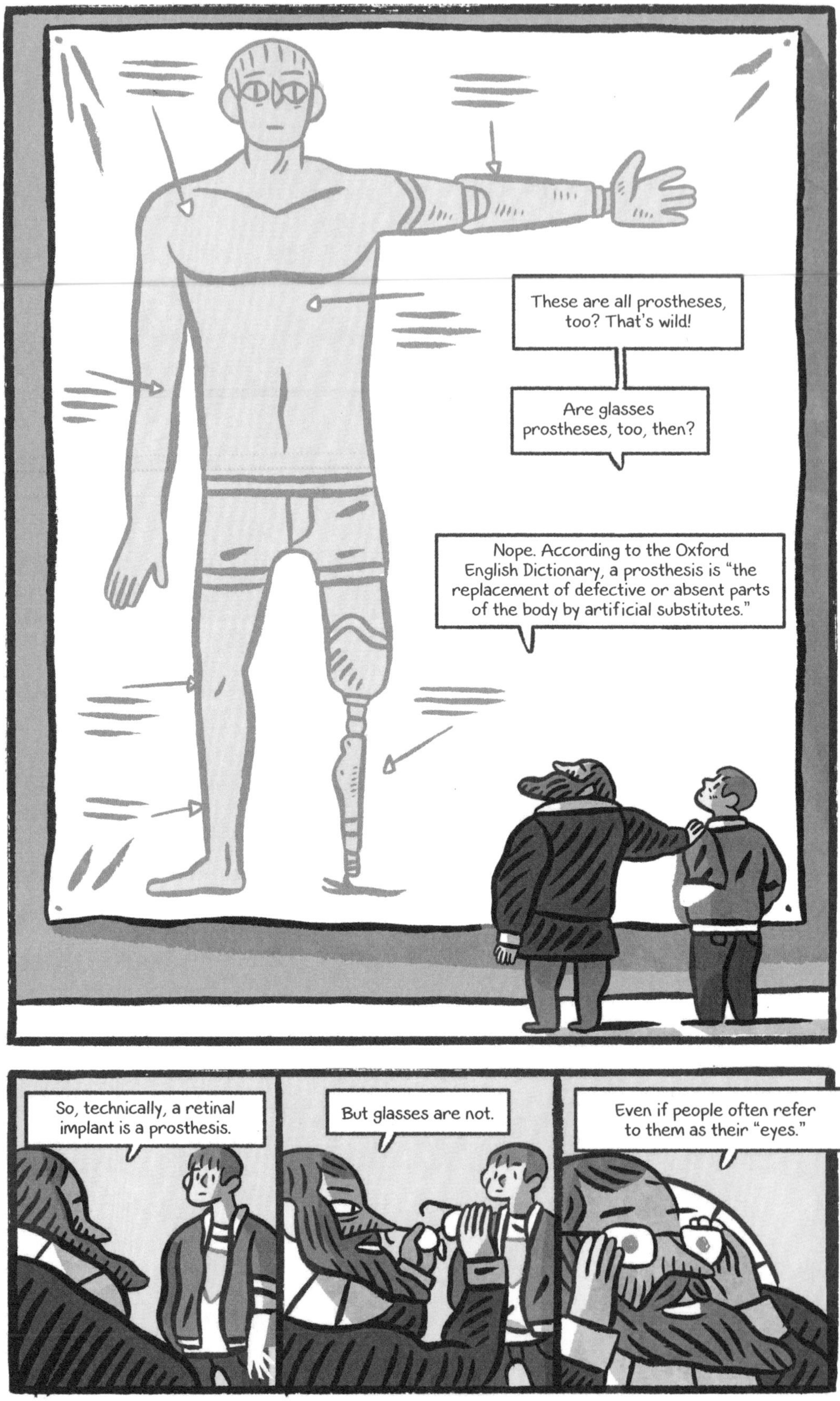
These are all prostheses, too? That's wild!
Are glasses prostheses, too, then?
Nope. According to the Oxford English Dictionary, a prosthesis is "the replacement of defective or absent parts of the body by artificial substitutes."
So, technically, a retinal implant is a prosthesis.
But glasses are not.
Even if people often refer to them as their "eyes."

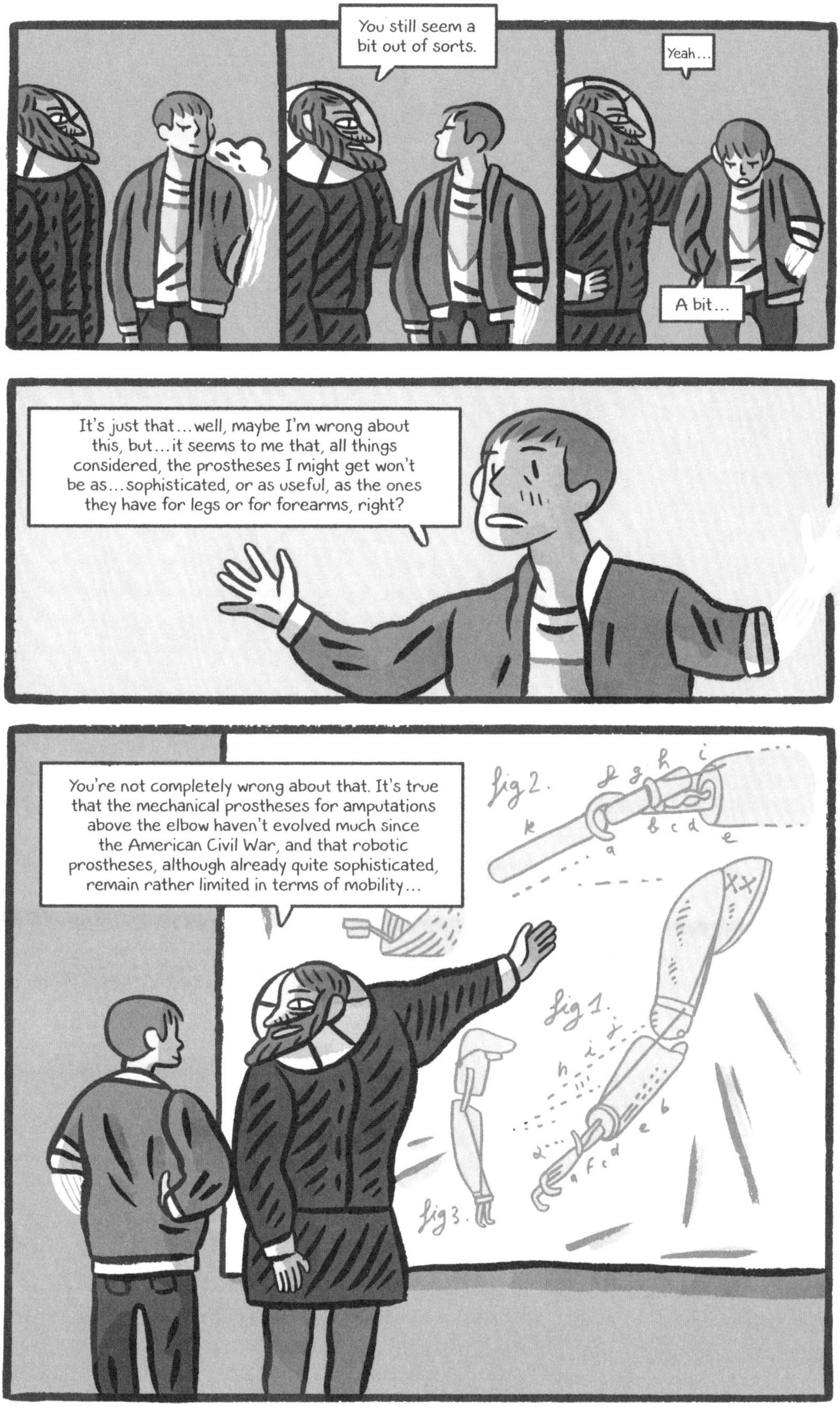
You still seem a bit out of sorts.
Yeah...
A bit...
It's just that...well, maybe I'm wrong about this, but...it seems to me that, all things considered, the prostheses I might get won't be as...sophisticated, or as useful, as the ones they have for legs or for forearms, right?
You're not completely wrong about that. It's true that the mechanical prostheses for amputations above the elbow haven't evolved much since the American Civil War, and that robotic prostheses, although already quite sophisticated, remain rather limited in terms of mobility...
fig 2.
fig 1.
fig 3.

because we use our muscles to make them work, and, in your case, all of the muscles in your hand and forearm were amputated. So, we only have use of your biceps and triceps.

There are only a few different degrees of freedom: one for the hand, one for the wrist, and one for the elbow. The commands are also very complicated. You have to "switch modes" for each joint.

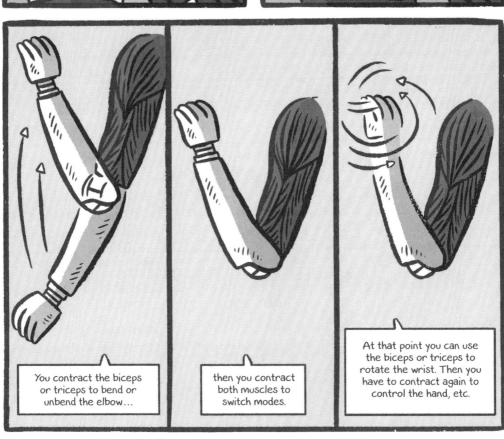

You contract the biceps or triceps to bend or unbend the elbow...

then you contract both muscles to switch modes.

At that point you can use the biceps or triceps to rotate the wrist. Then you have to contract again to control the hand, etc.

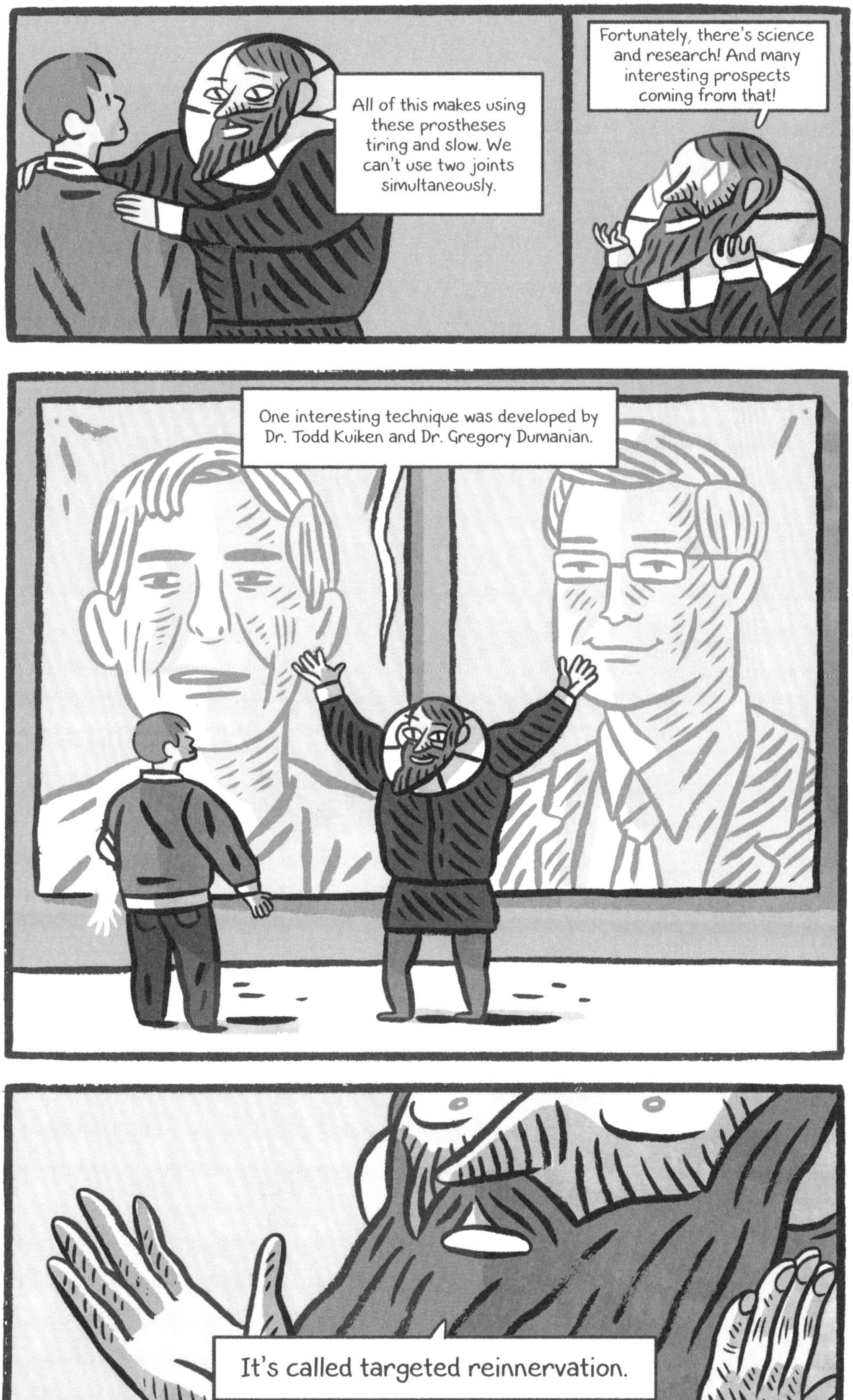
All of this makes using these prostheses tiring and slow. We can't use two joints simultaneously.
Fortunately, there's science and research! And many interesting prospects coming from that!
One interesting technique was developed by Dr. Todd Kuiken and Dr. Gregory Dumanian.
It's called targeted reinnervation.

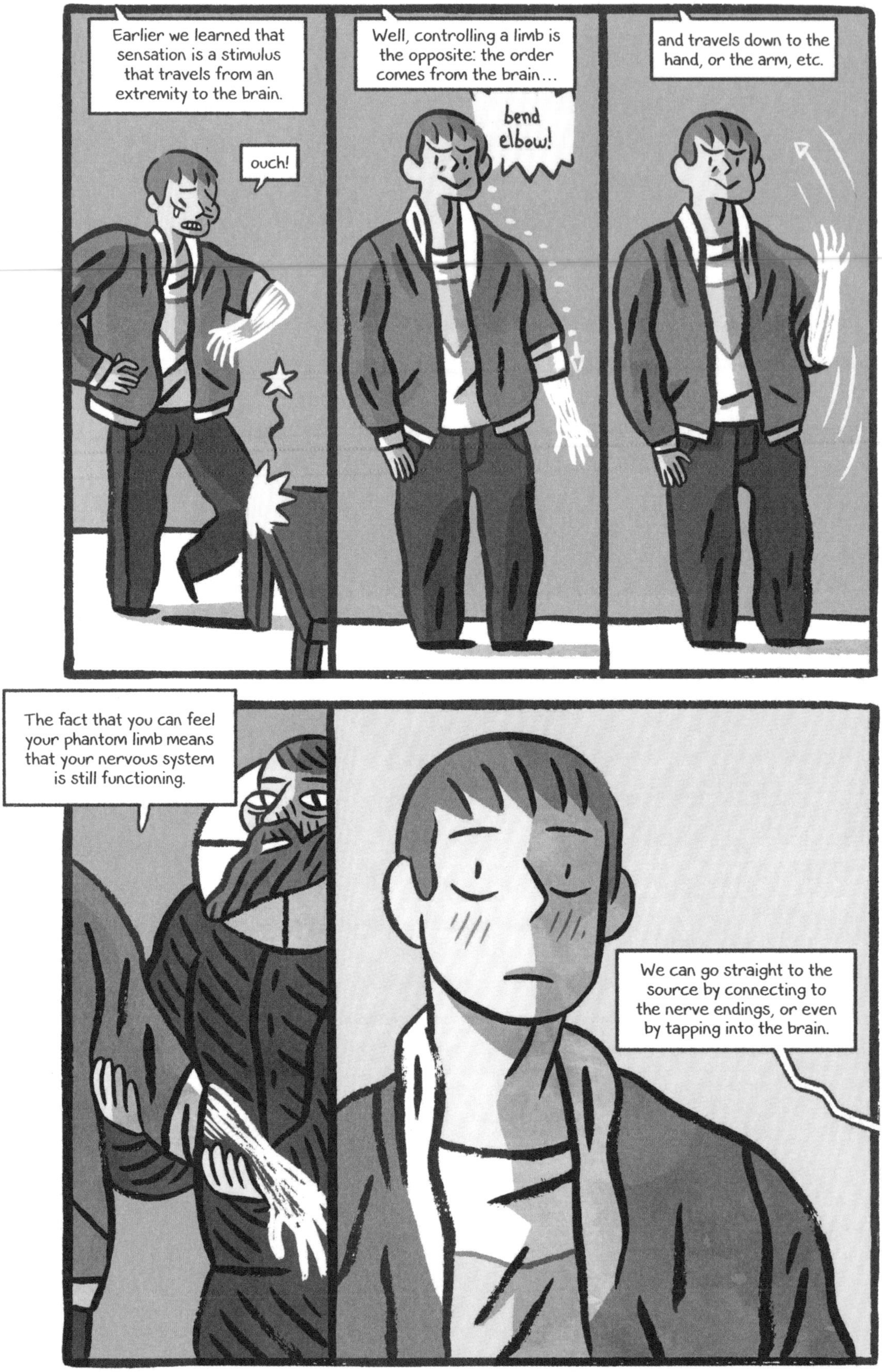
Earlier we learned that sensation is a stimulus that travels from an extremity to the brain.
ouch!
Well, controlling a limb is the opposite: the order comes from the brain…
bend elbow!
and travels down to the hand, or the arm, etc.
The fact that you can feel your phantom limb means that your nervous system is still functioning.
We can go straight to the source by connecting to the nerve endings, or even by tapping into the brain.

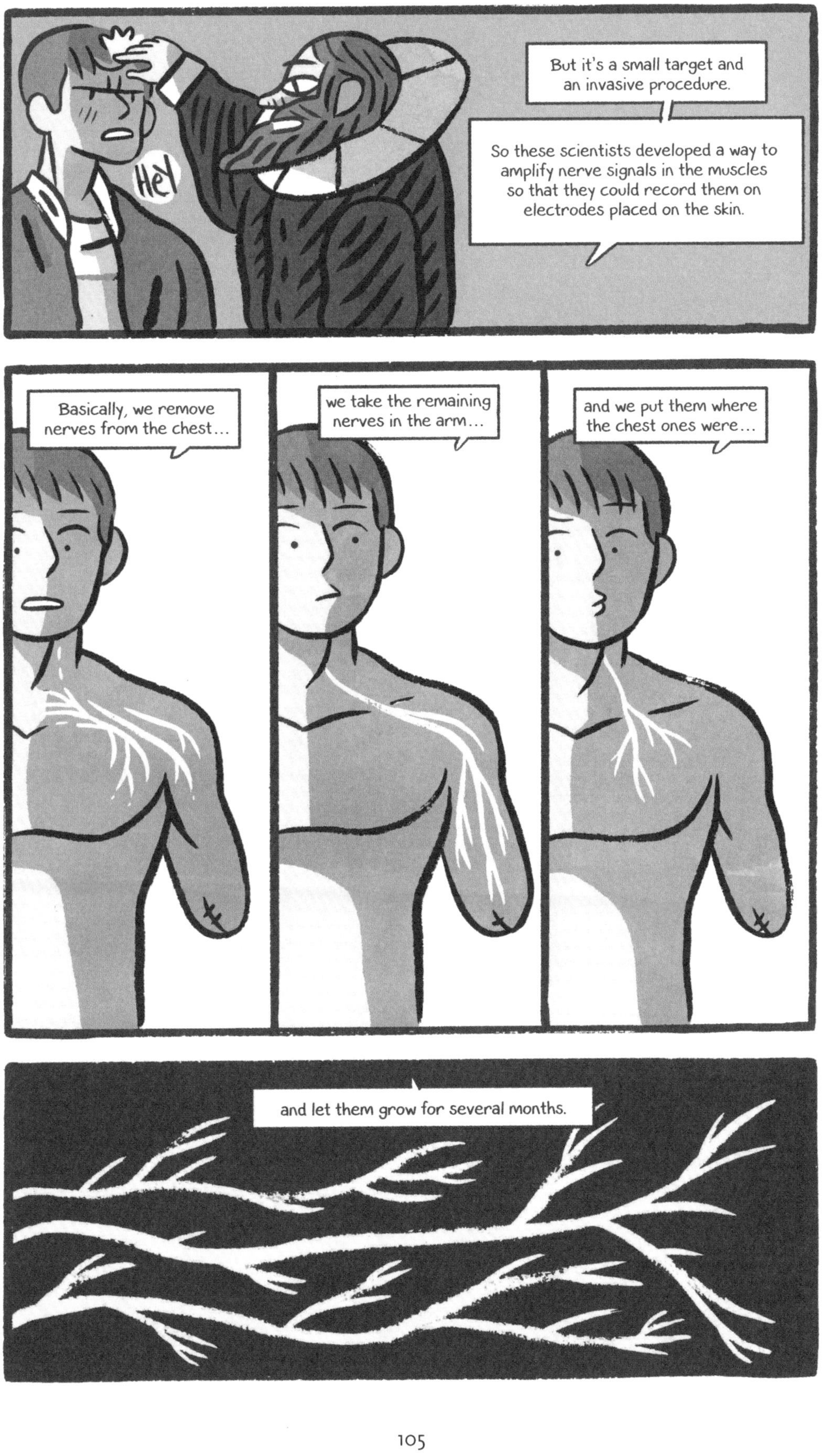
But it's a small target and an invasive procedure.
HEY
So these scientists developed a way to amplify nerve signals in the muscles so that they could record them on electrodes placed on the skin.
Basically, we remove nerves from the chest…
we take the remaining nerves in the arm…
and we put them where the chest ones were…
and let them grow for several months.

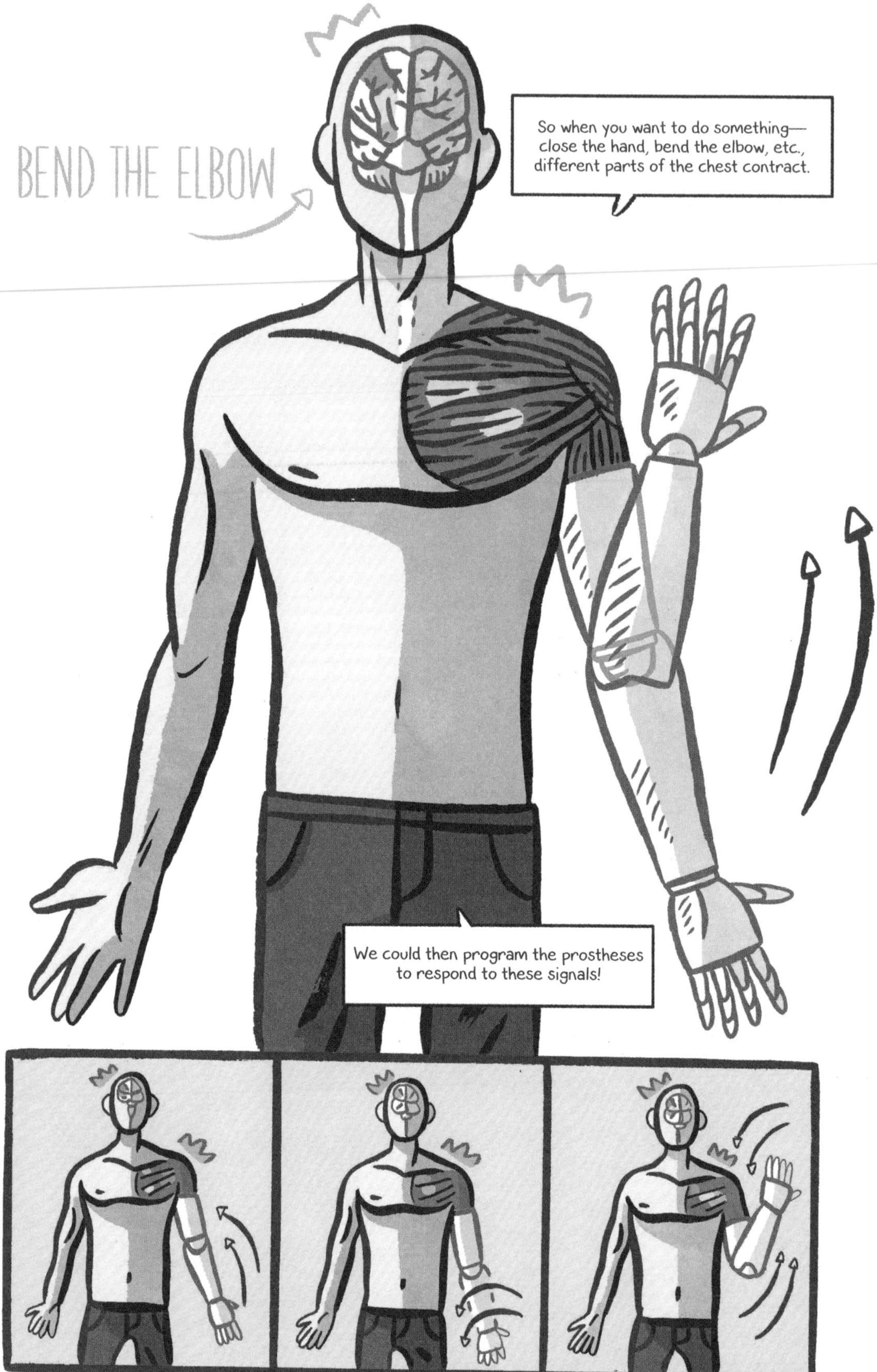
BEND THE ELBOW
So when you want to do something—close the hand, bend the elbow, etc., different parts of the chest contract.
We could then program the prostheses to respond to these signals!

These prostheses would be more intuitive to use because you wouldn't have to consciously think "contract this or that muscle." Your brain could just order the action. They would also be faster, smoother, and would allow the use of several joints at once!

The research isn't finished, but we can hope to one day be able to bend our fingers individually, use joints to their full potential...and even feel things!

Feel things?

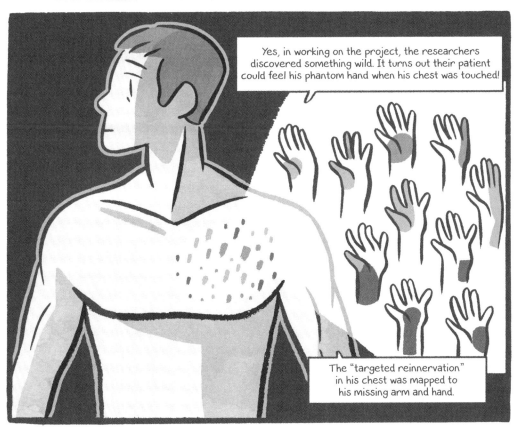

Yes, in working on the project, the researchers discovered something wild. It turns out their patient could feel his phantom hand when his chest was touched!

The "targeted reinnervation" in his chest was mapped to his missing arm and hand.

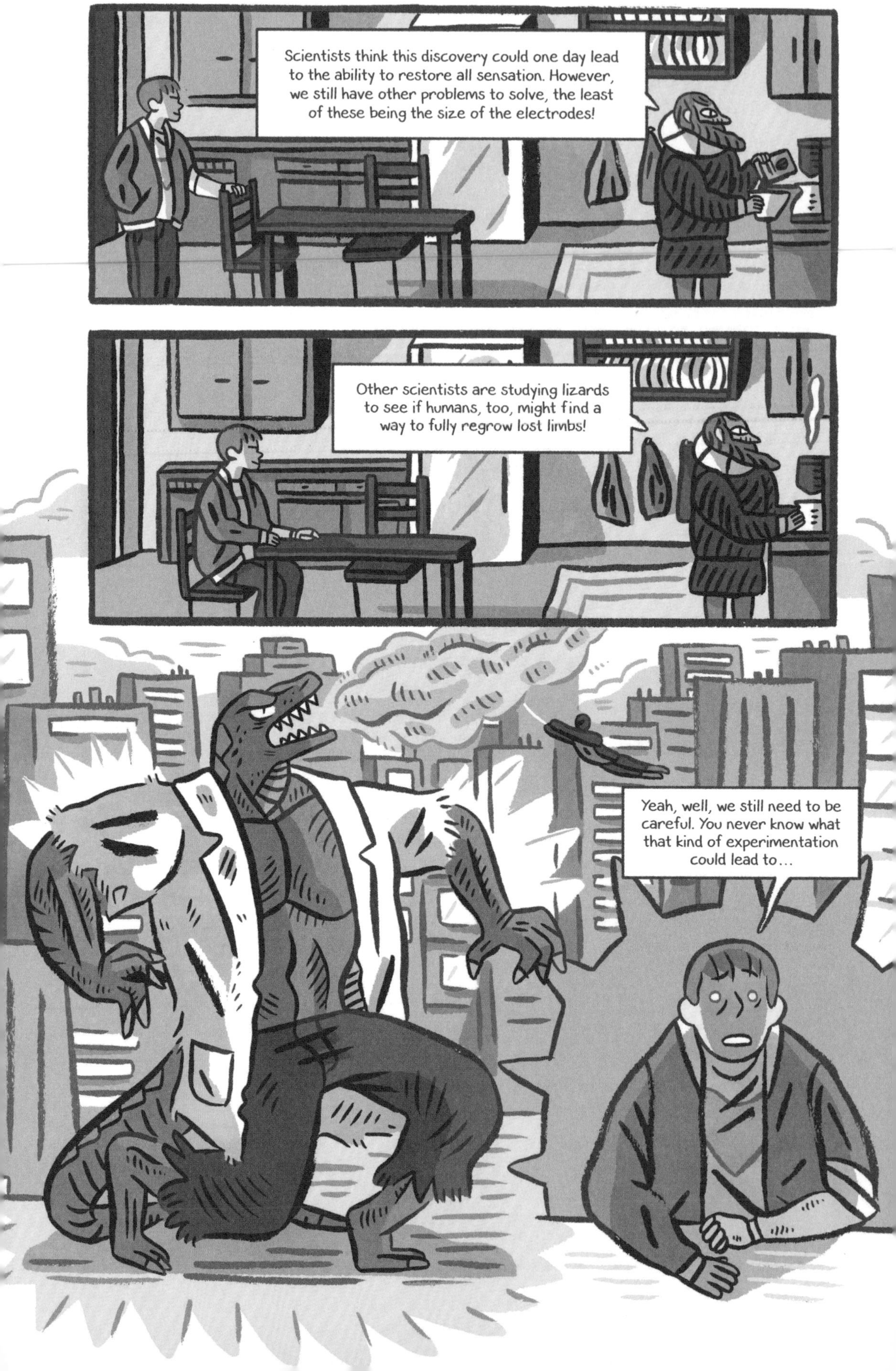
Scientists think this discovery could one day lead to the ability to restore all sensation. However, we still have other problems to solve, the least of these being the size of the electrodes!
Other scientists are studying lizards to see if humans, too, might find a way to fully regrow lost limbs!
Yeah, well, we still need to be careful. You never know what that kind of experimentation could lead to...

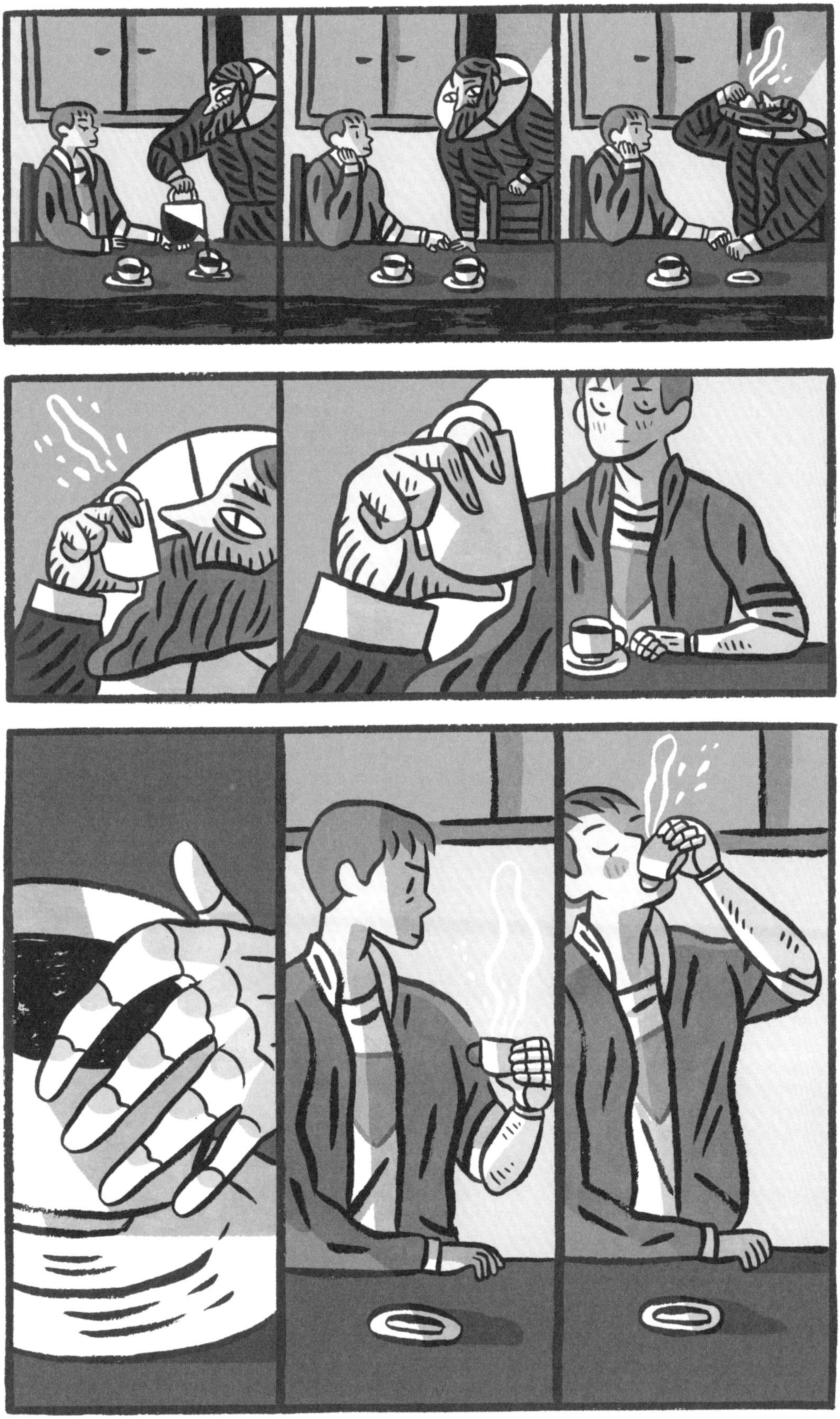

BRIEF INTERLUDE – ARTICULATED PROSTHESES
(BY AMBROISE PARÉ)

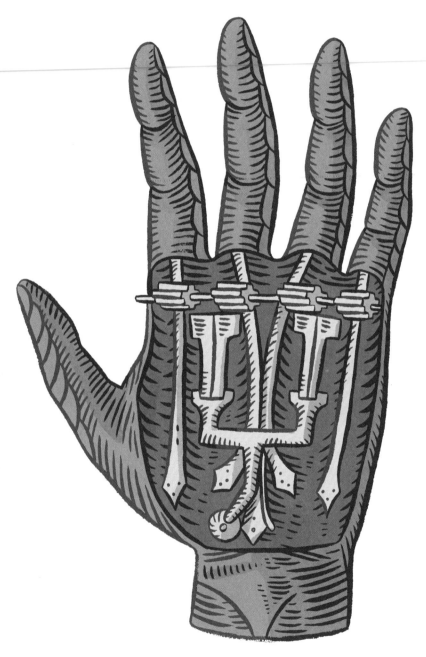

Artificial hand designed by Ambroise Paré, allowing for
articulation at the joints and having the appearance of a real hand.

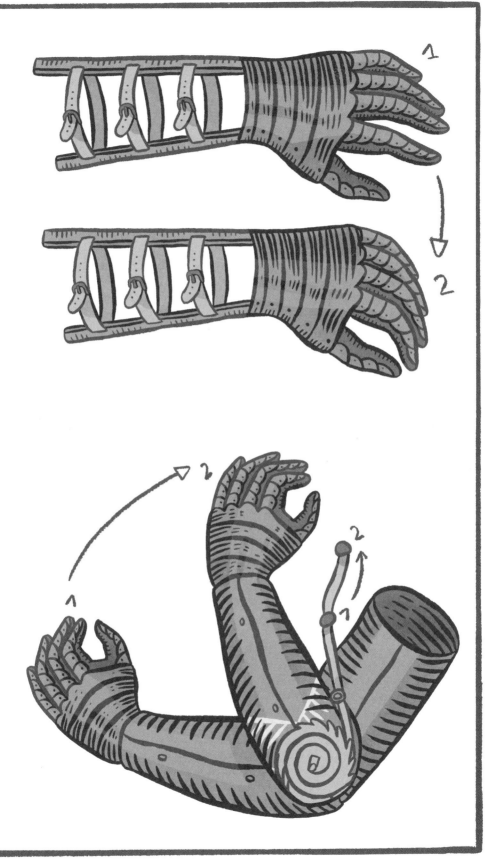

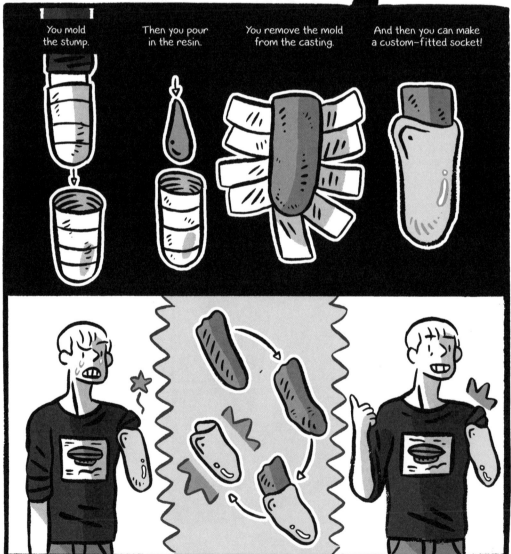

You mold
the stump.

Then you pour
in the resin.

You remove the mold
from the casting.

And then you can make
a custom-fitted socket!

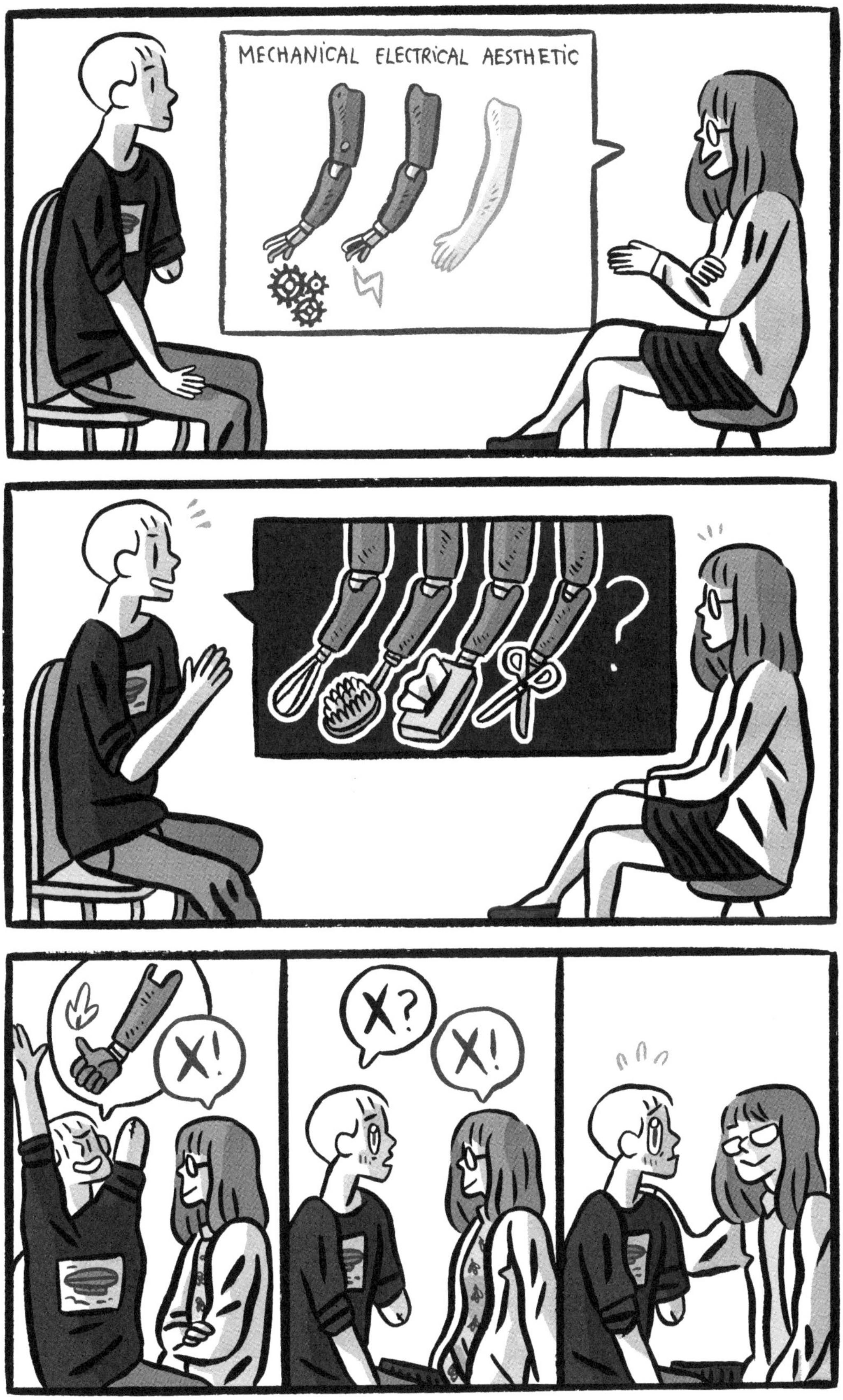
MECHANICAL ELECTRICAL AESTHETIC
?
X!
X?
X!

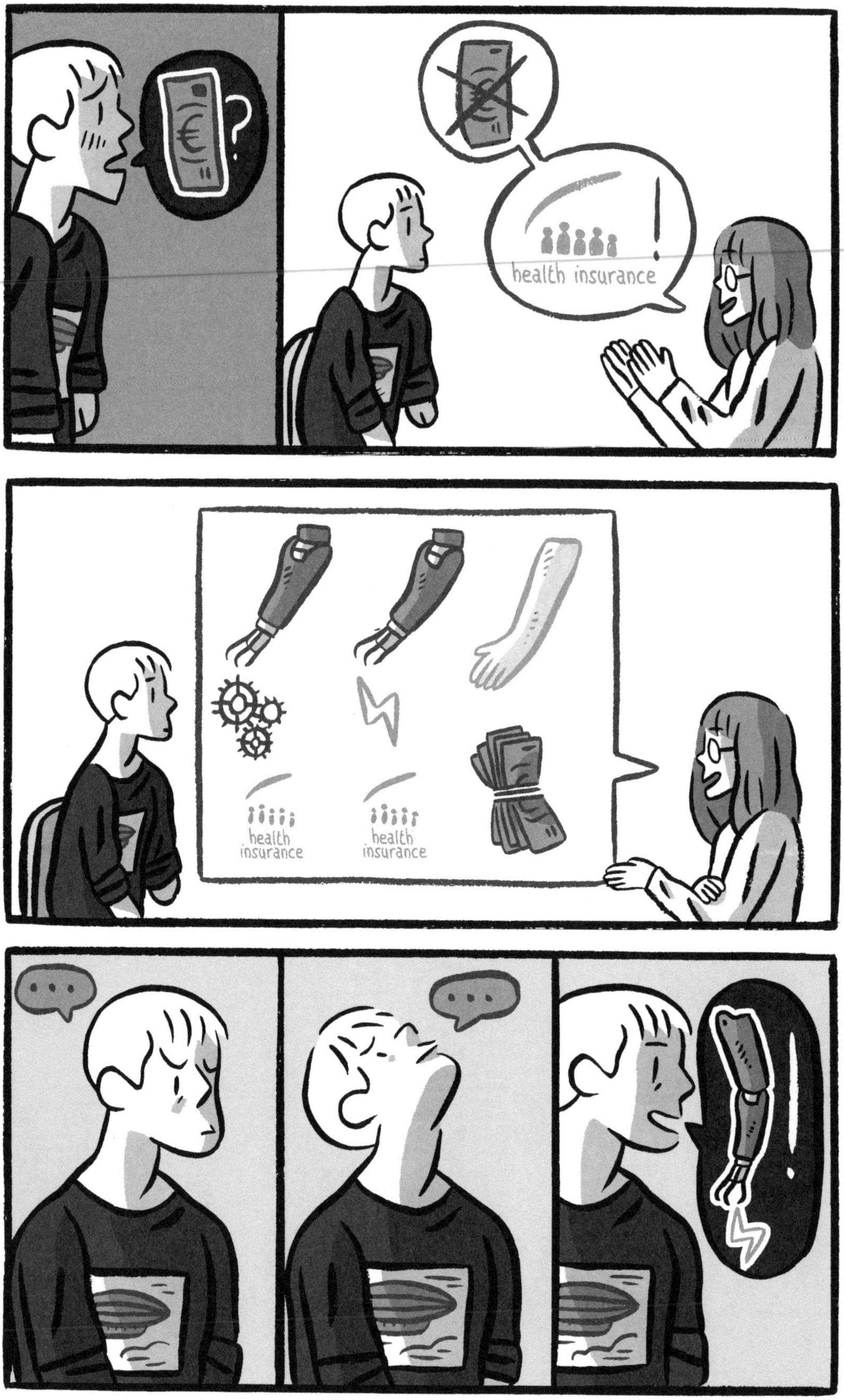
?
health insurance
!
health insurance
health insurance

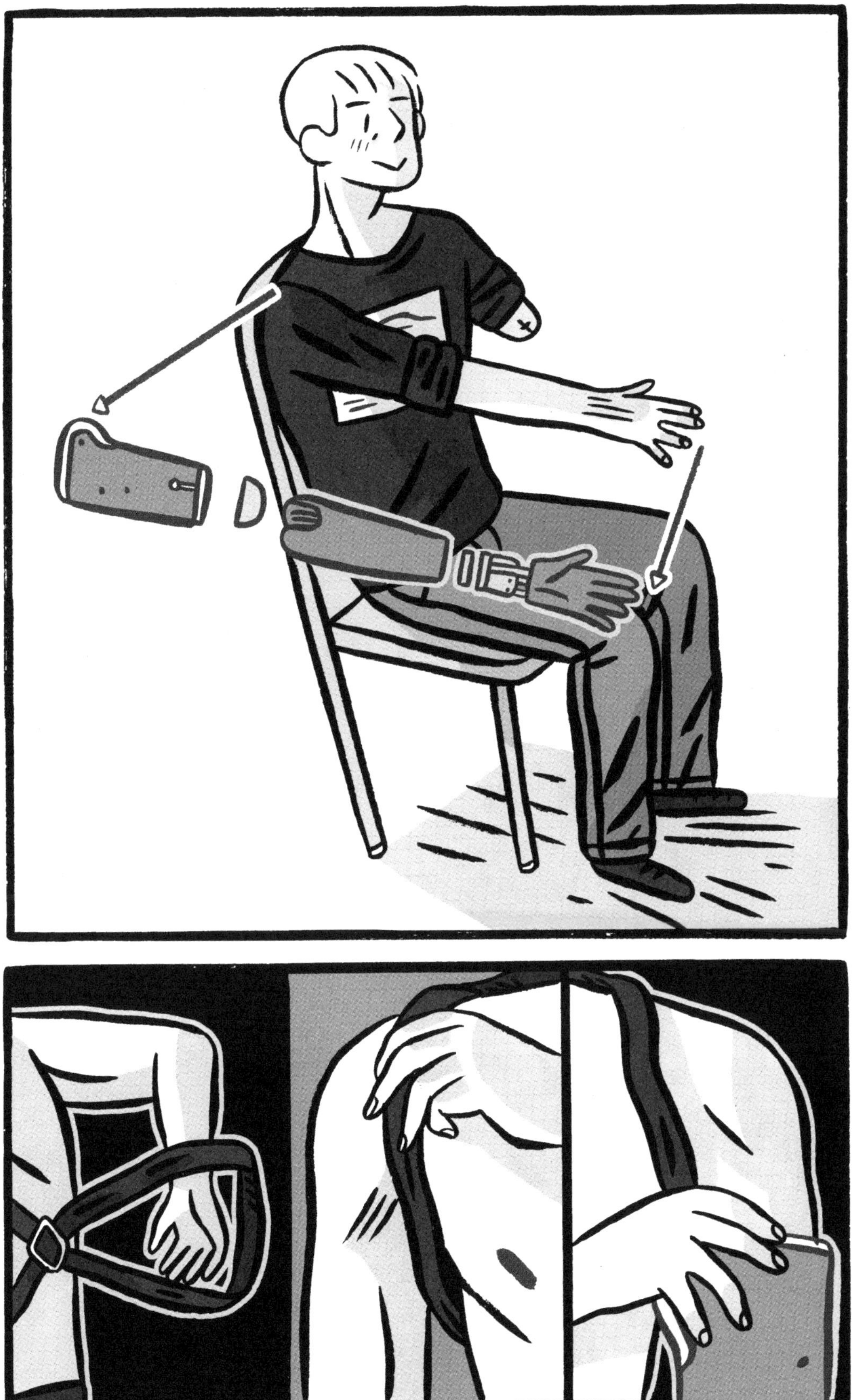

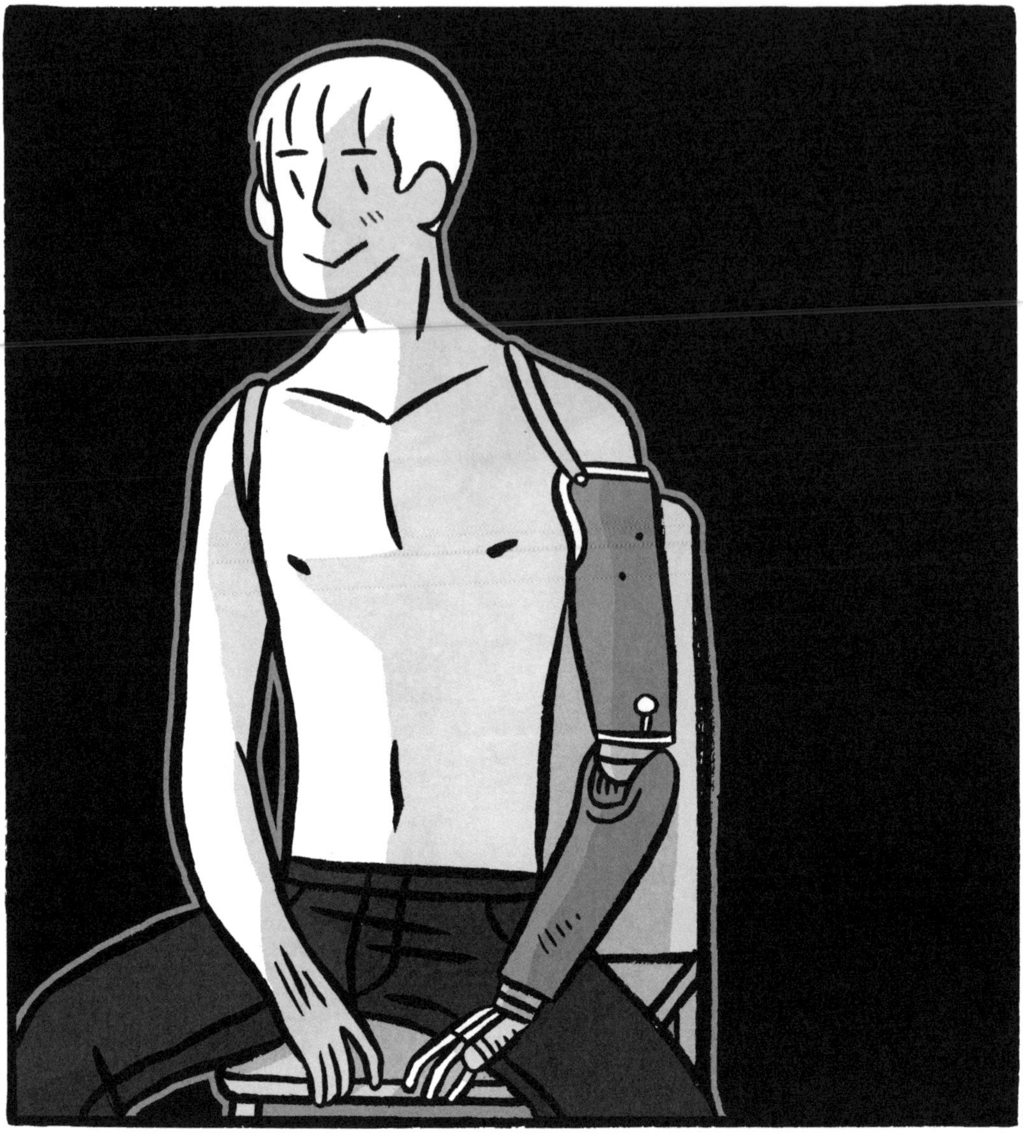

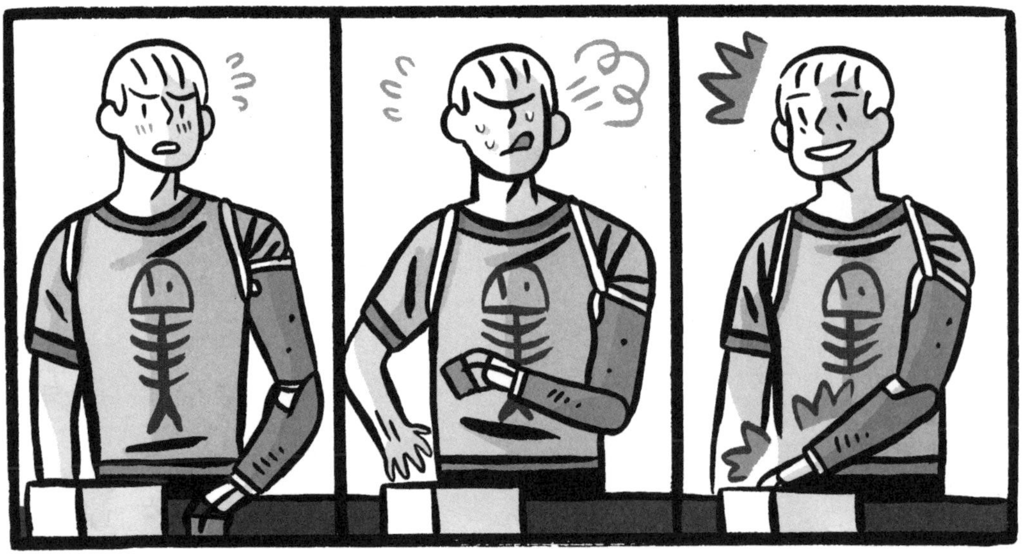

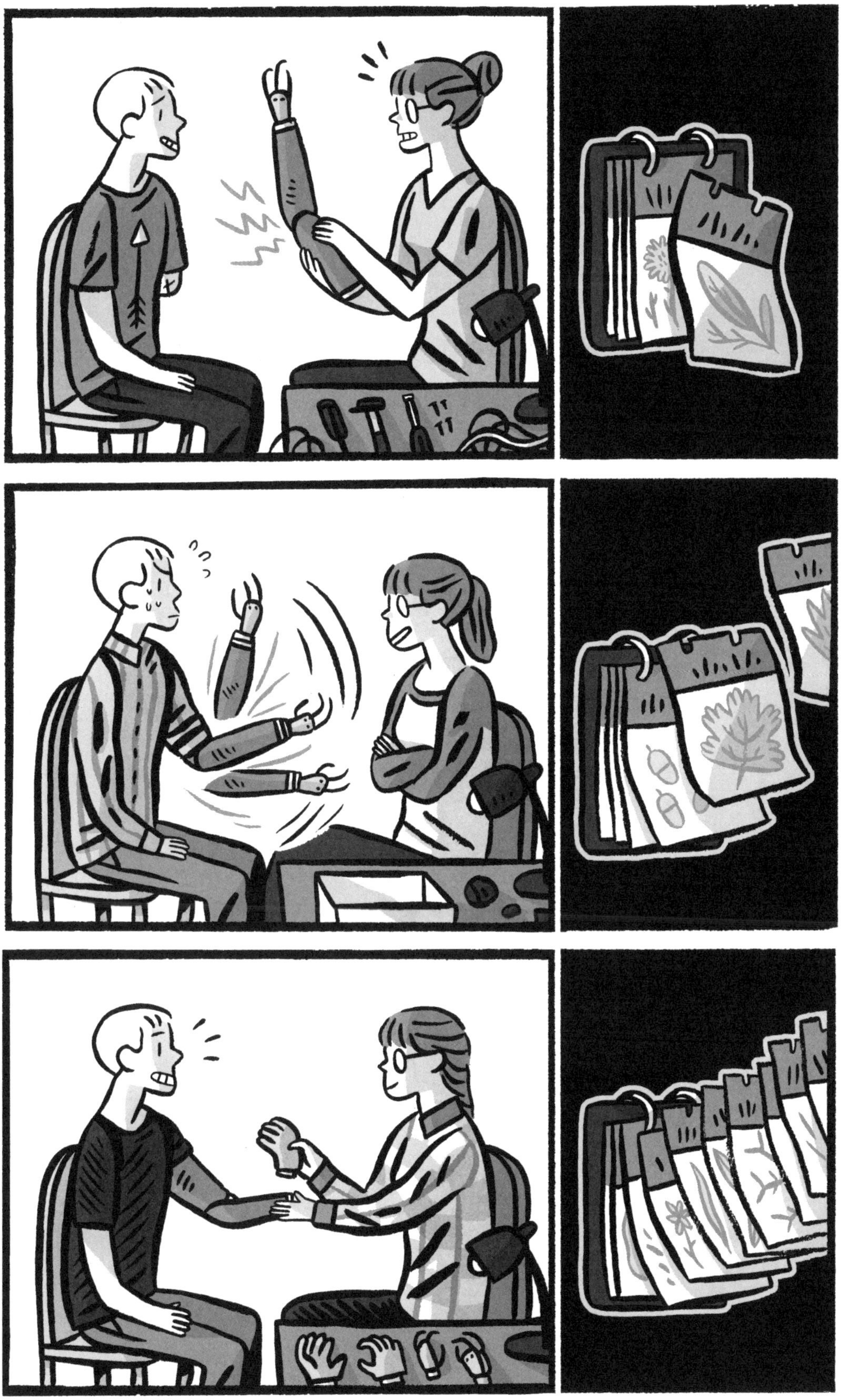

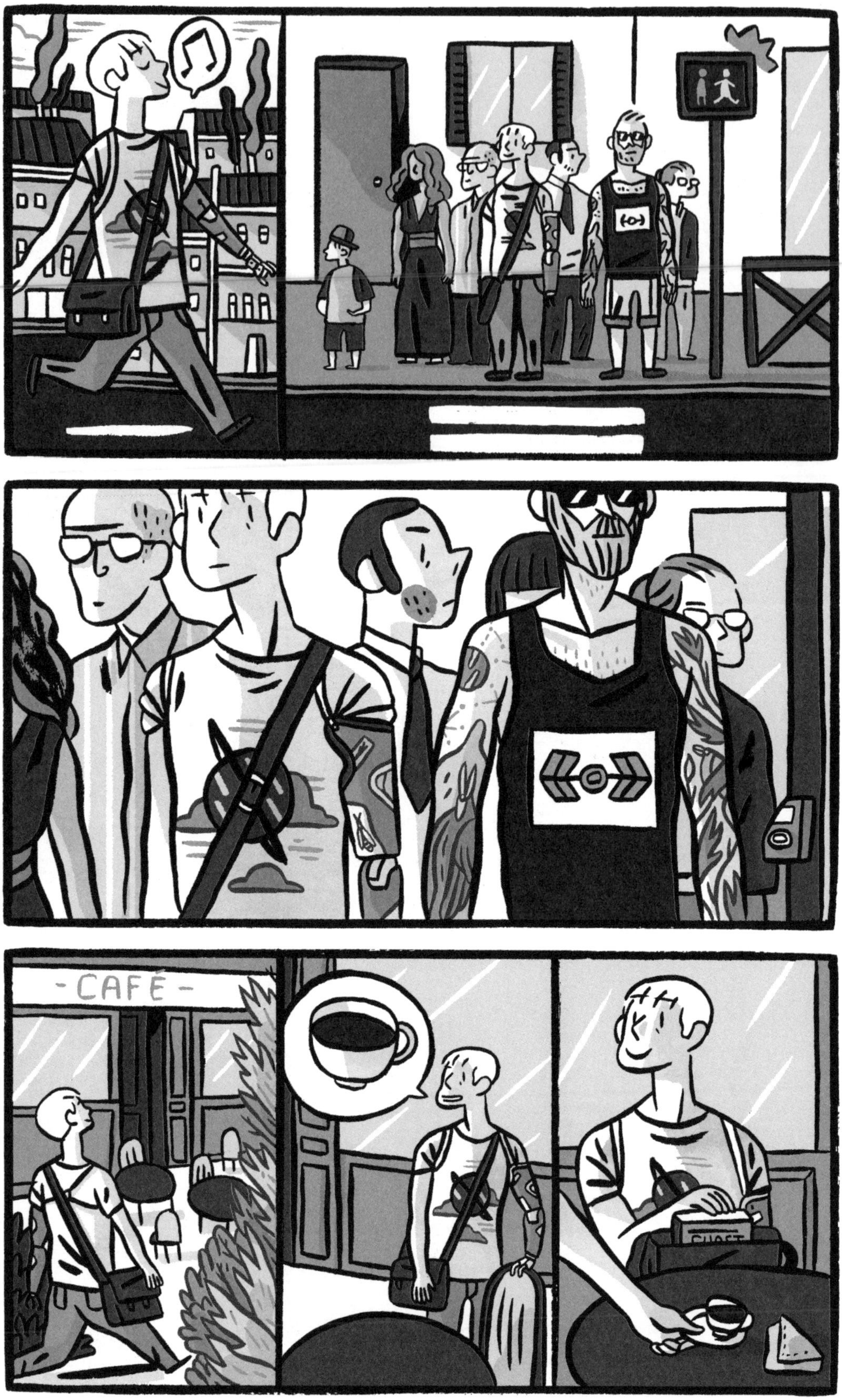
-CAFÉ-

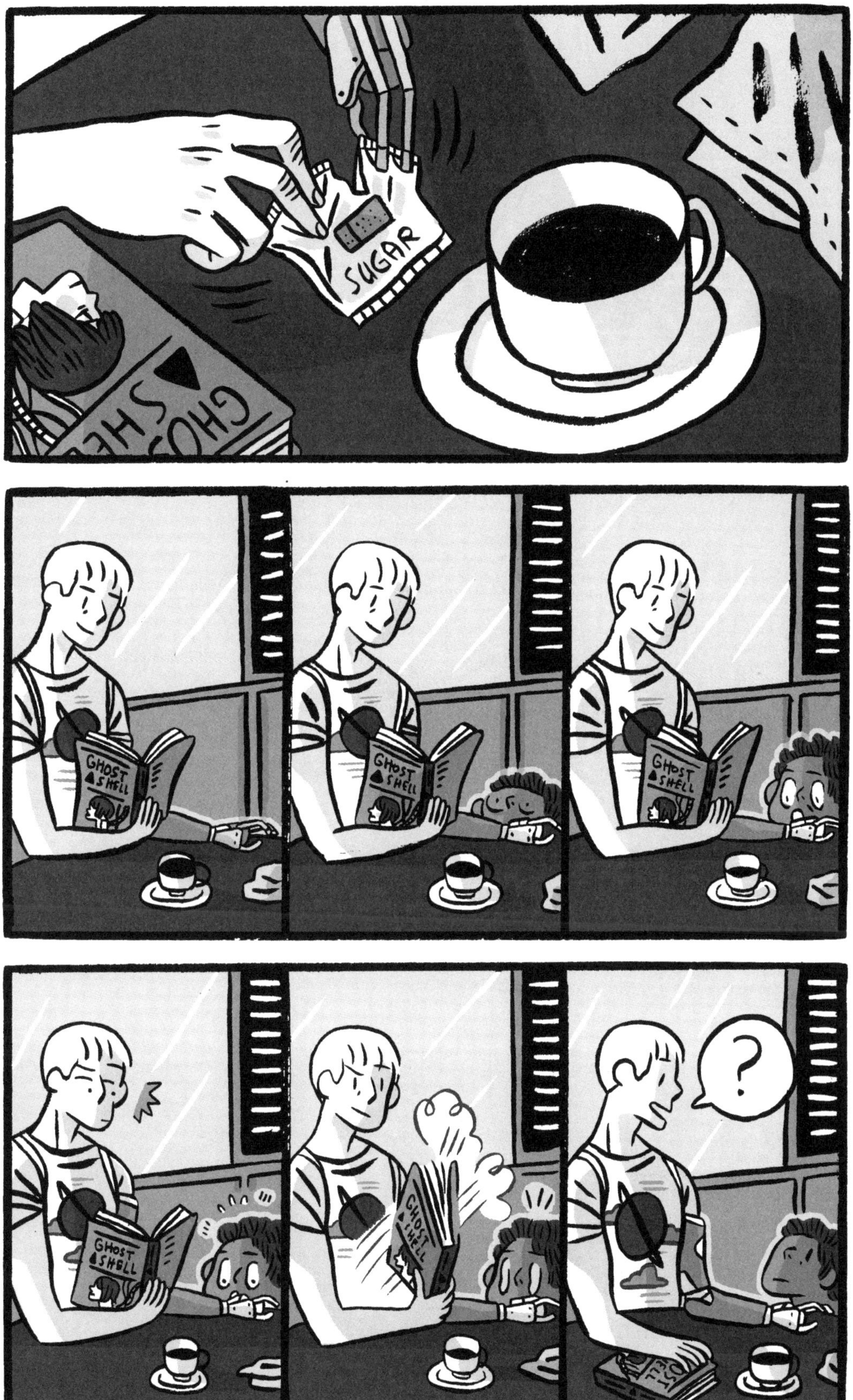
SUGAR
GHOST SHELL
?

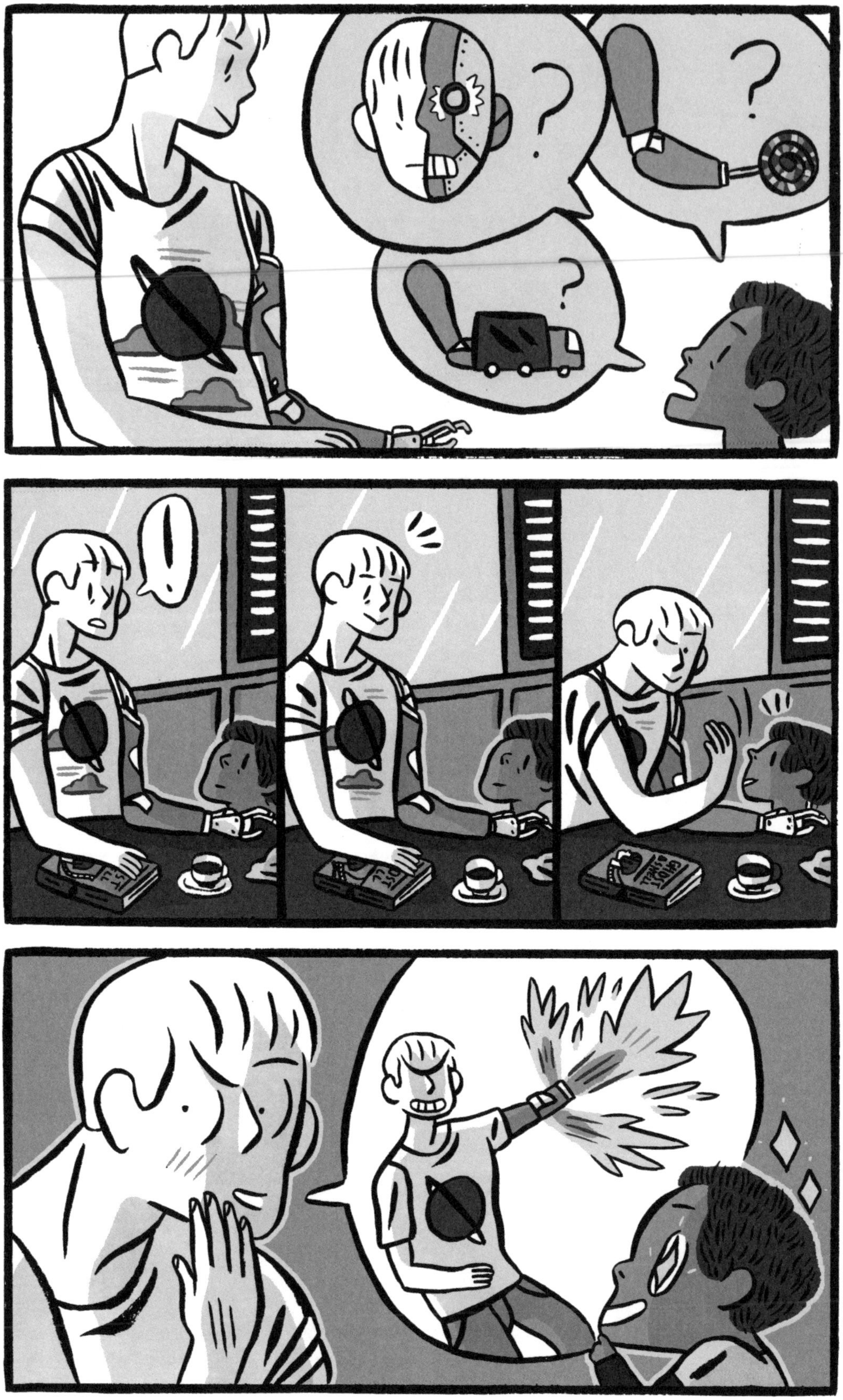

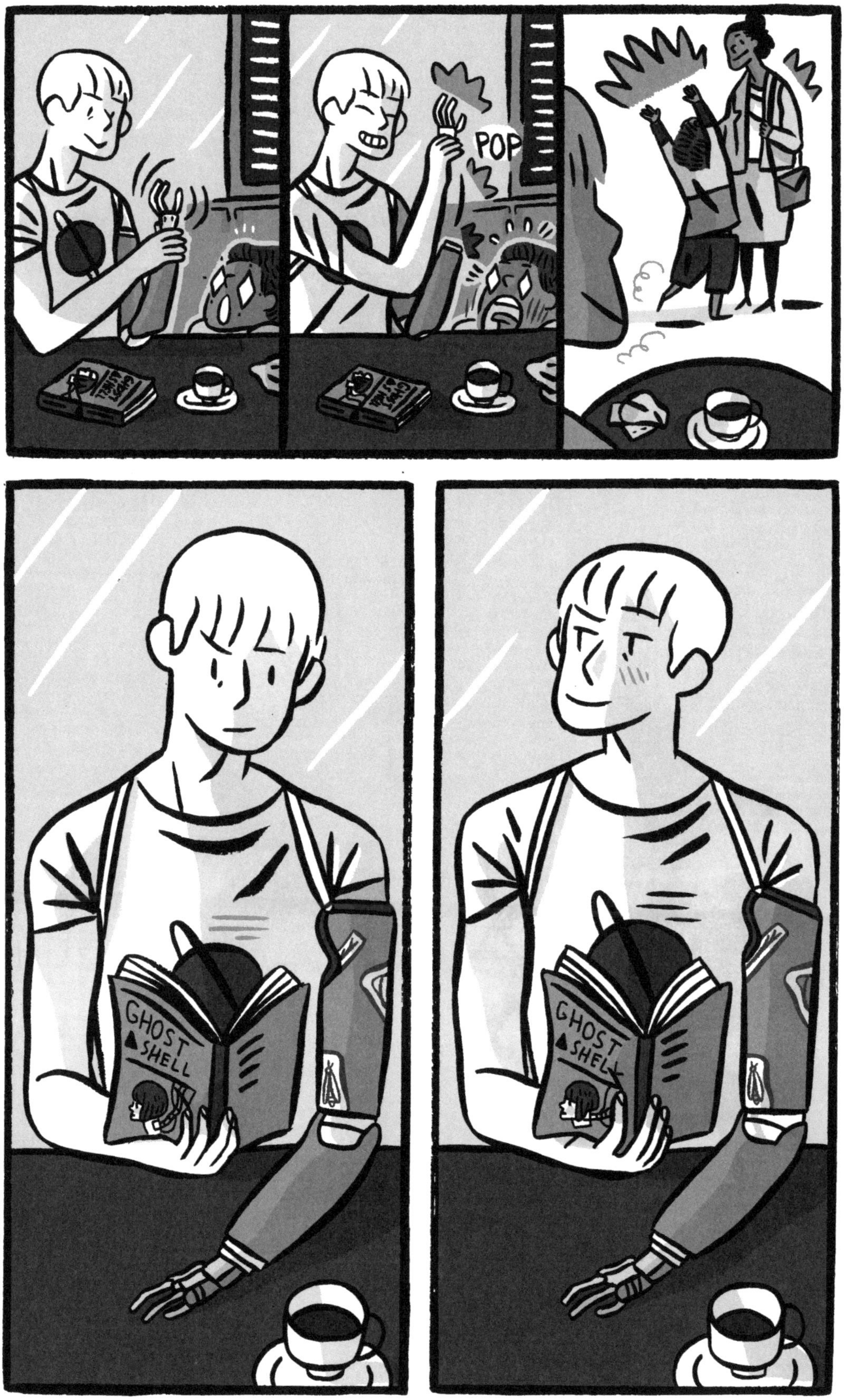
POP
GHOST SHELL
GHOST SHELL

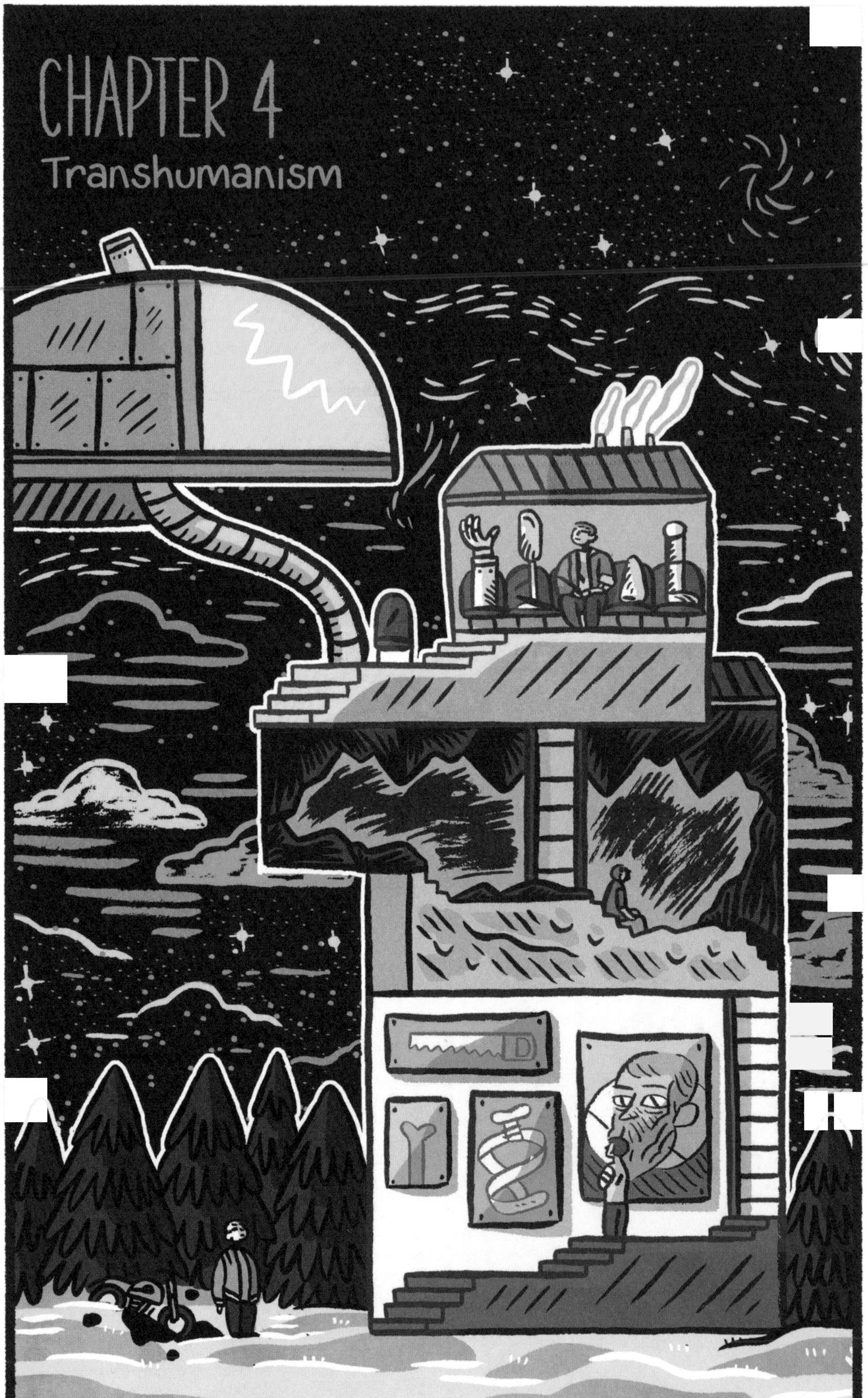
CHAPTER 4
Transhumanism

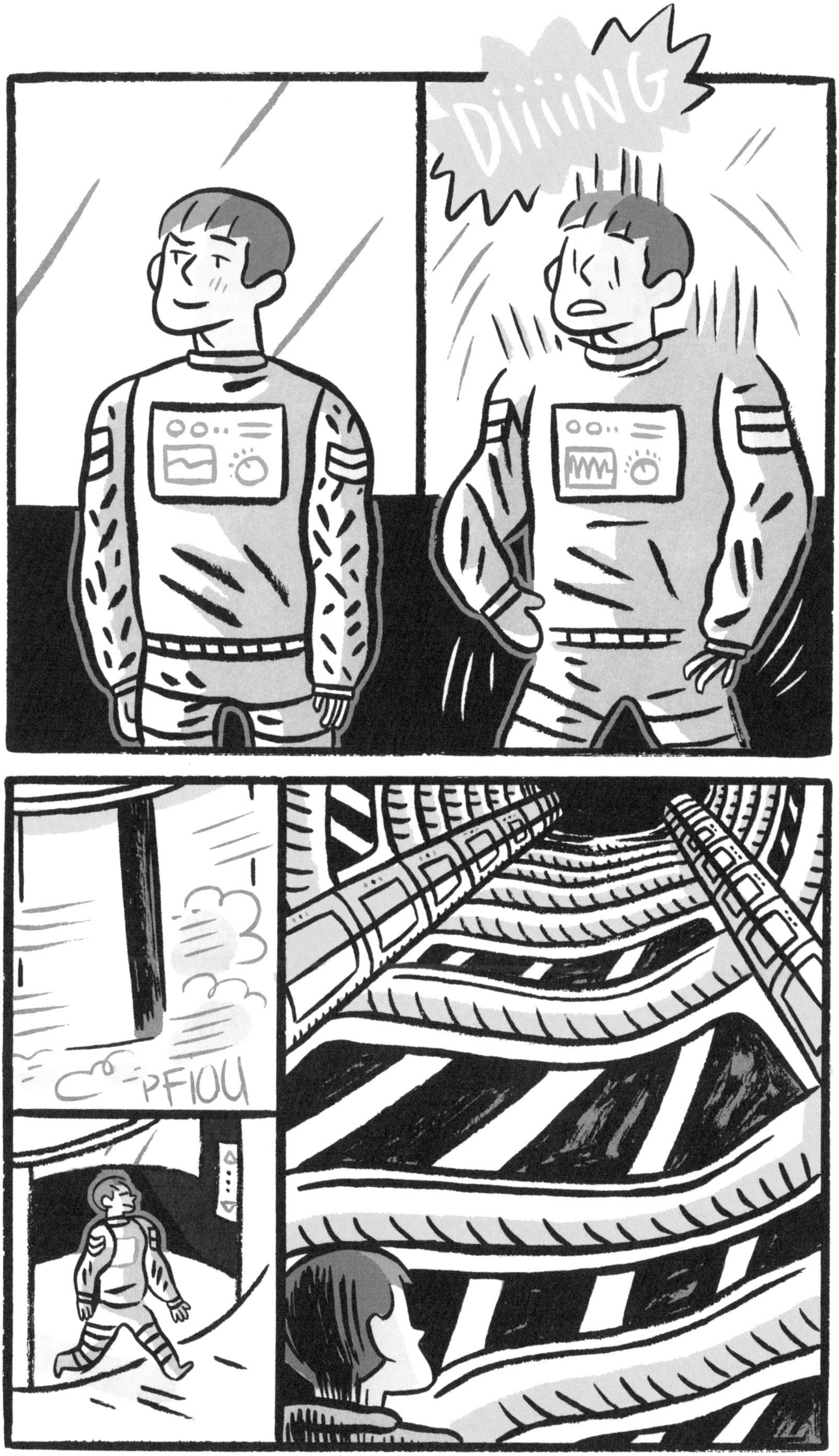
DIIIING
PFIOU

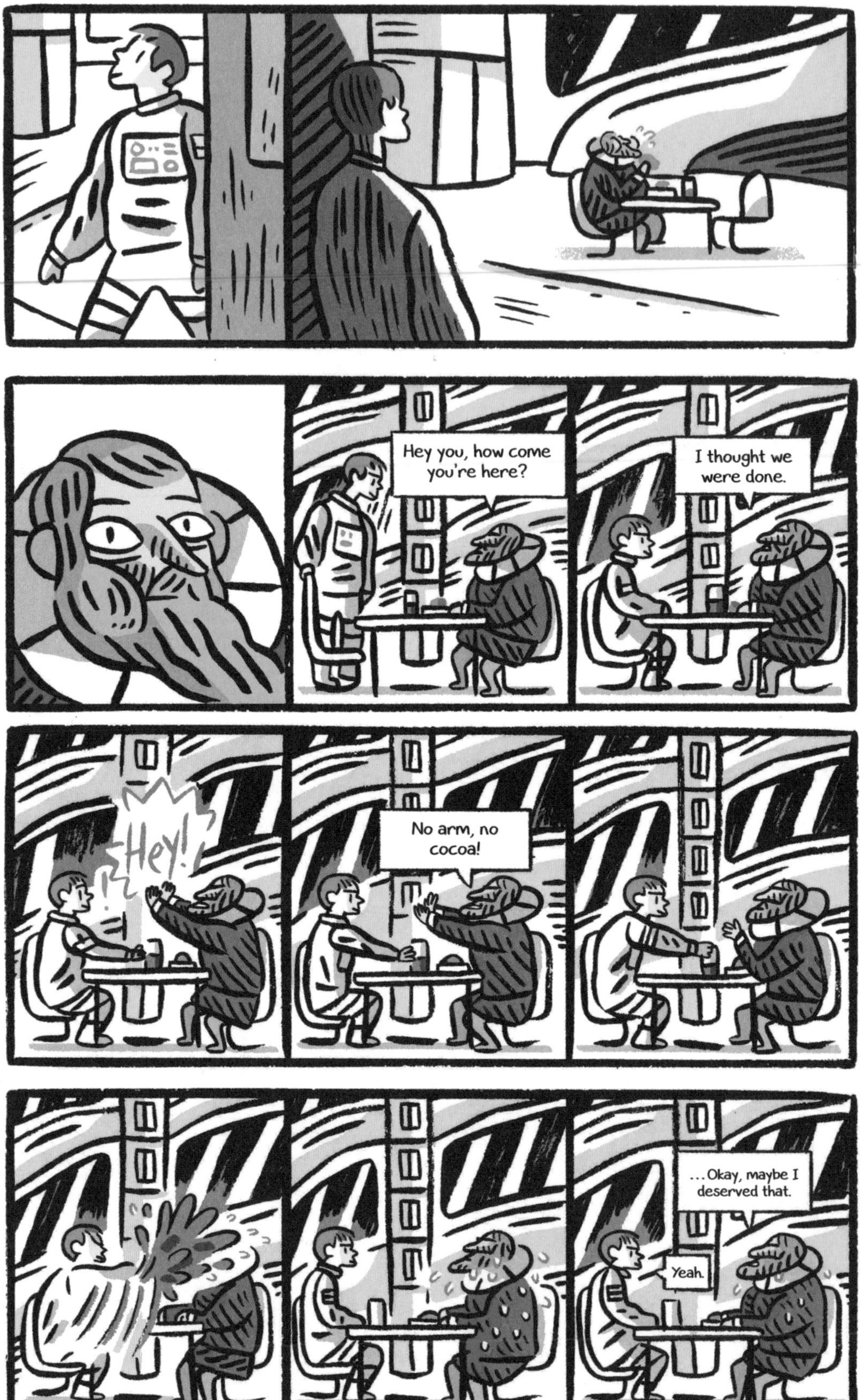
Hey you, how come you're here?
I thought we were done.
Hey!
!
No arm, no cocoa!
...Okay, maybe I deserved that.
Yeah.

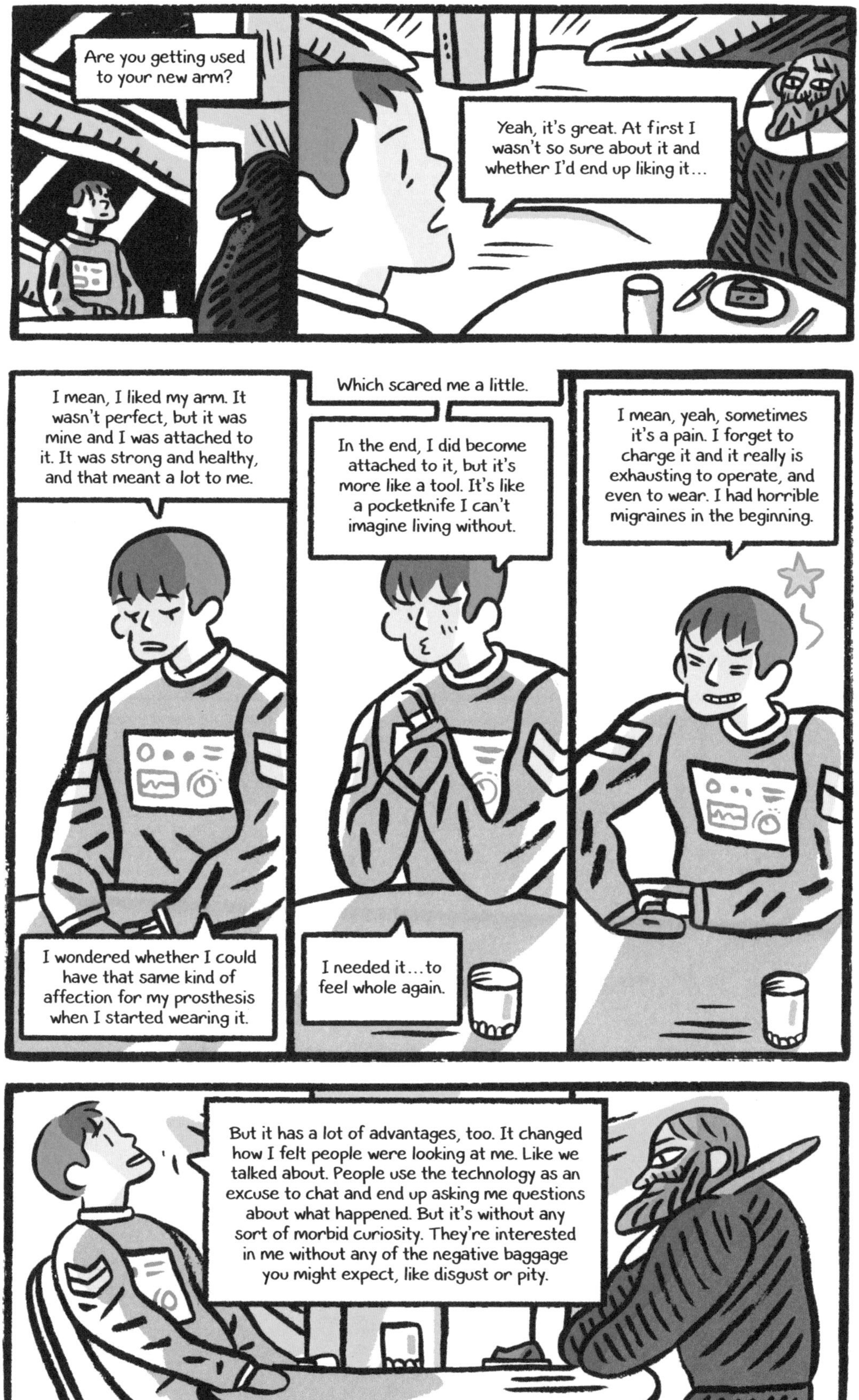
Are you getting used to your new arm?
Yeah, it's great. At first I wasn't so sure about it and whether I'd end up liking it...
I mean, I liked my arm. It wasn't perfect, but it was mine and I was attached to it. It was strong and healthy, and that meant a lot to me.
I wondered whether I could have that same kind of affection for my prosthesis when I started wearing it.
Which scared me a little.
In the end, I did become attached to it, but it's more like a tool. It's like a pocketknife I can't imagine living without.
I needed it...to feel whole again.
I mean, yeah, sometimes it's a pain. I forget to charge it and it really is exhausting to operate, and even to wear. I had horrible migraines in the beginning.
But it has a lot of advantages, too. It changed how I felt people were looking at me. Like we talked about. People use the technology as an excuse to chat and end up asking me questions about what happened. But it's without any sort of morbid curiosity. They're interested in me without any of the negative baggage you might expect, like disgust or pity.

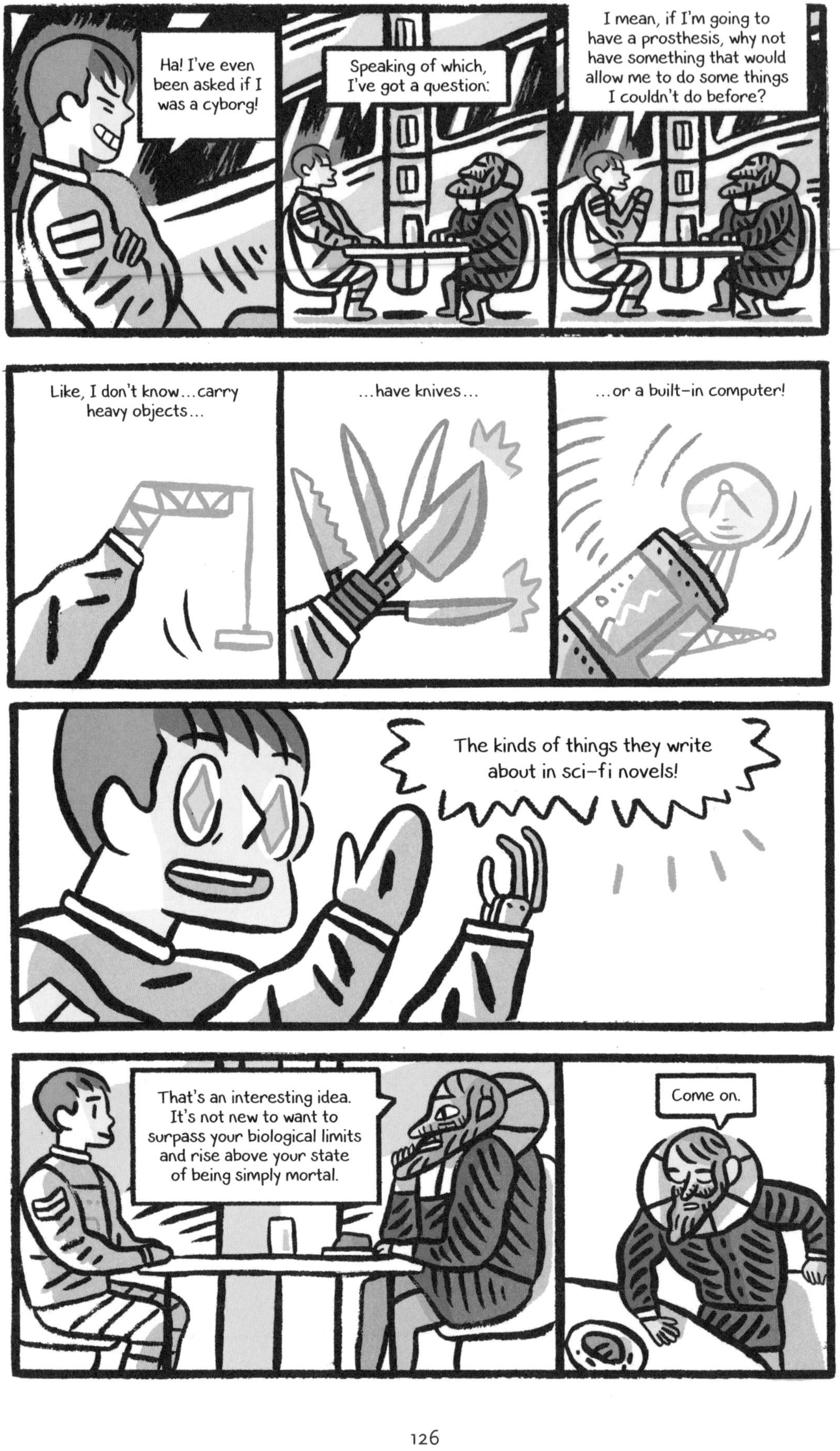
Ha! I've even been asked if I was a cyborg!
Speaking of which, I've got a question:
I mean, if I'm going to have a prosthesis, why not have something that would allow me to do some things I couldn't do before?
Like, I don't know...carry heavy objects...
...have knives...
...or a built-in computer!
The kinds of things they write about in sci-fi novels!
That's an interesting idea. It's not new to want to surpass your biological limits and rise above your state of being simply mortal.
Come on.

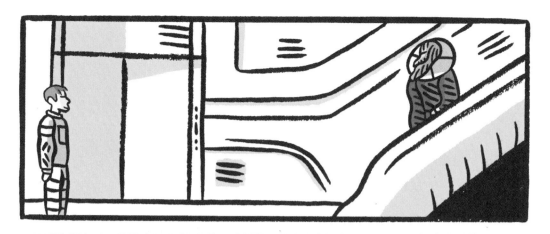

Consider Icarus, who built wings so that he could fly! Or Prometheus, who stole fire, the symbol of knowledge, and offered it to humans, allowing them to rise above their condition, threatening the very supremacy of the gods. We know how that ended: not well at all!

Thanks to the rise of technology over the last few centuries, the ability not only to repair and reconstruct ourselves, but also to augment ourselves, is becoming more of a reality, and it's leading to new ways of thinking.

Do you see where this is going?

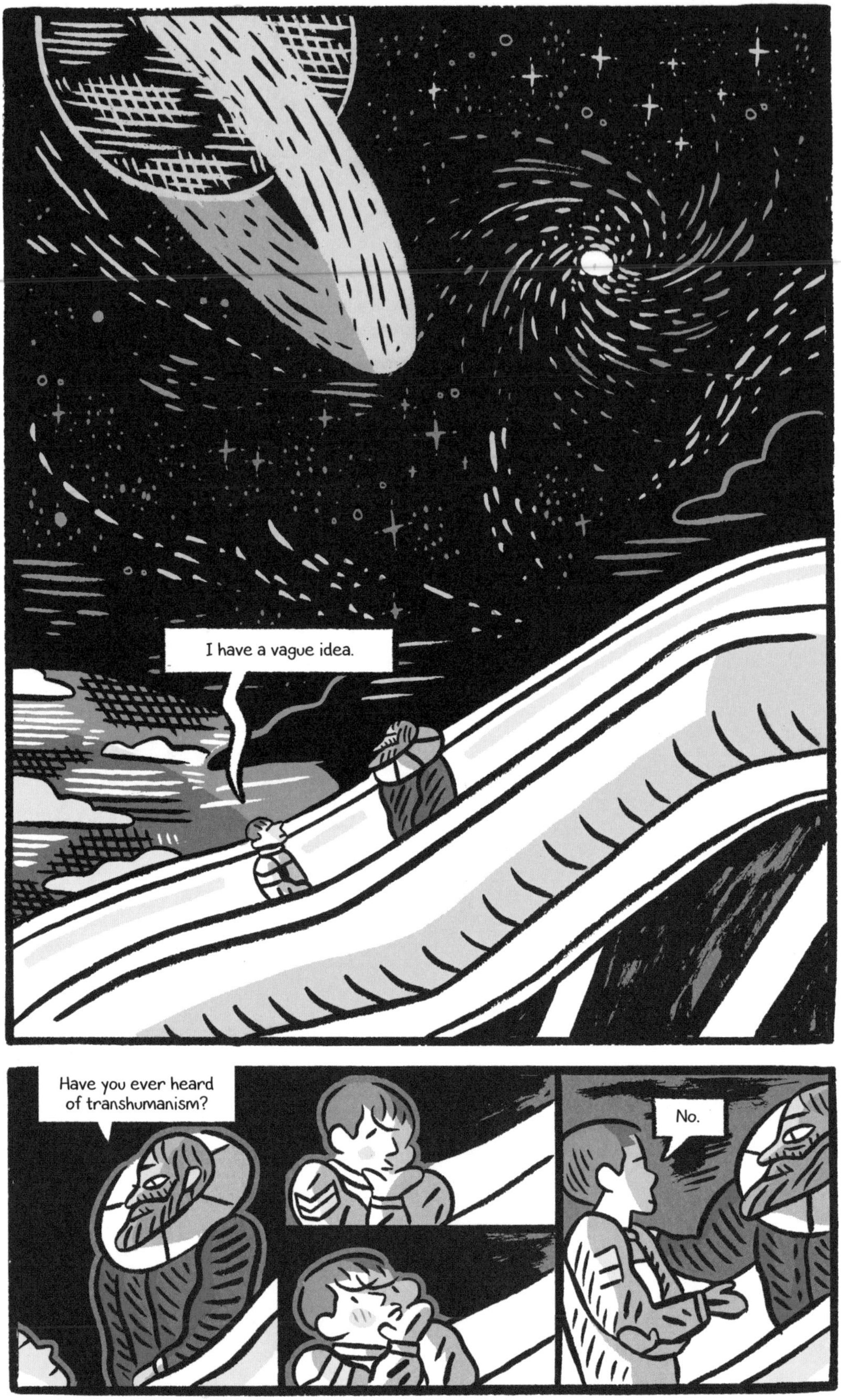
I have a vague idea.
Have you ever heard of transhumanism?
No.

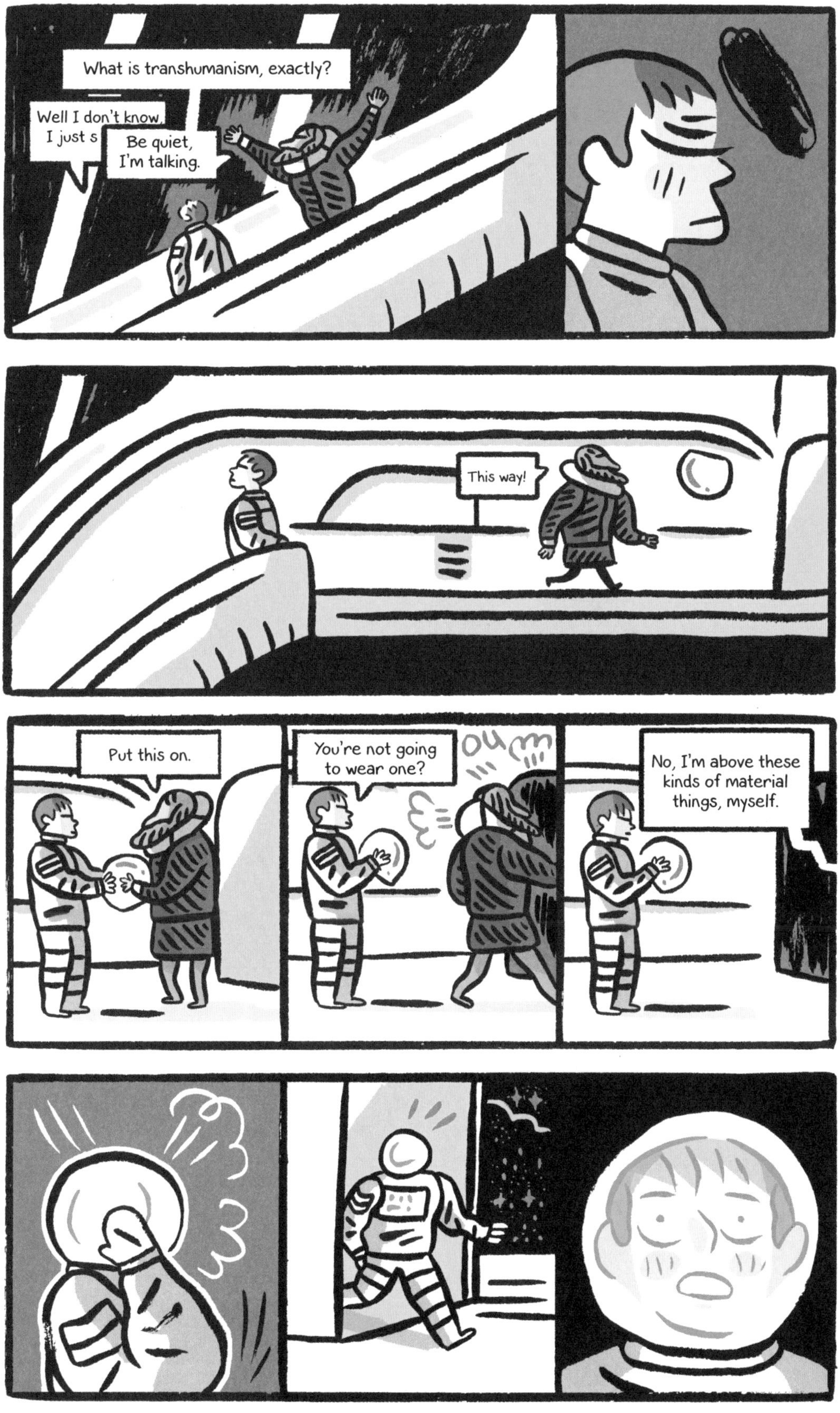
What is transhumanism, exactly?
Well I don't know, I just s
Be quiet, I'm talking.
This way!
Put this on.
You're not going to wear one?
No, I'm above these kinds of material things, myself.

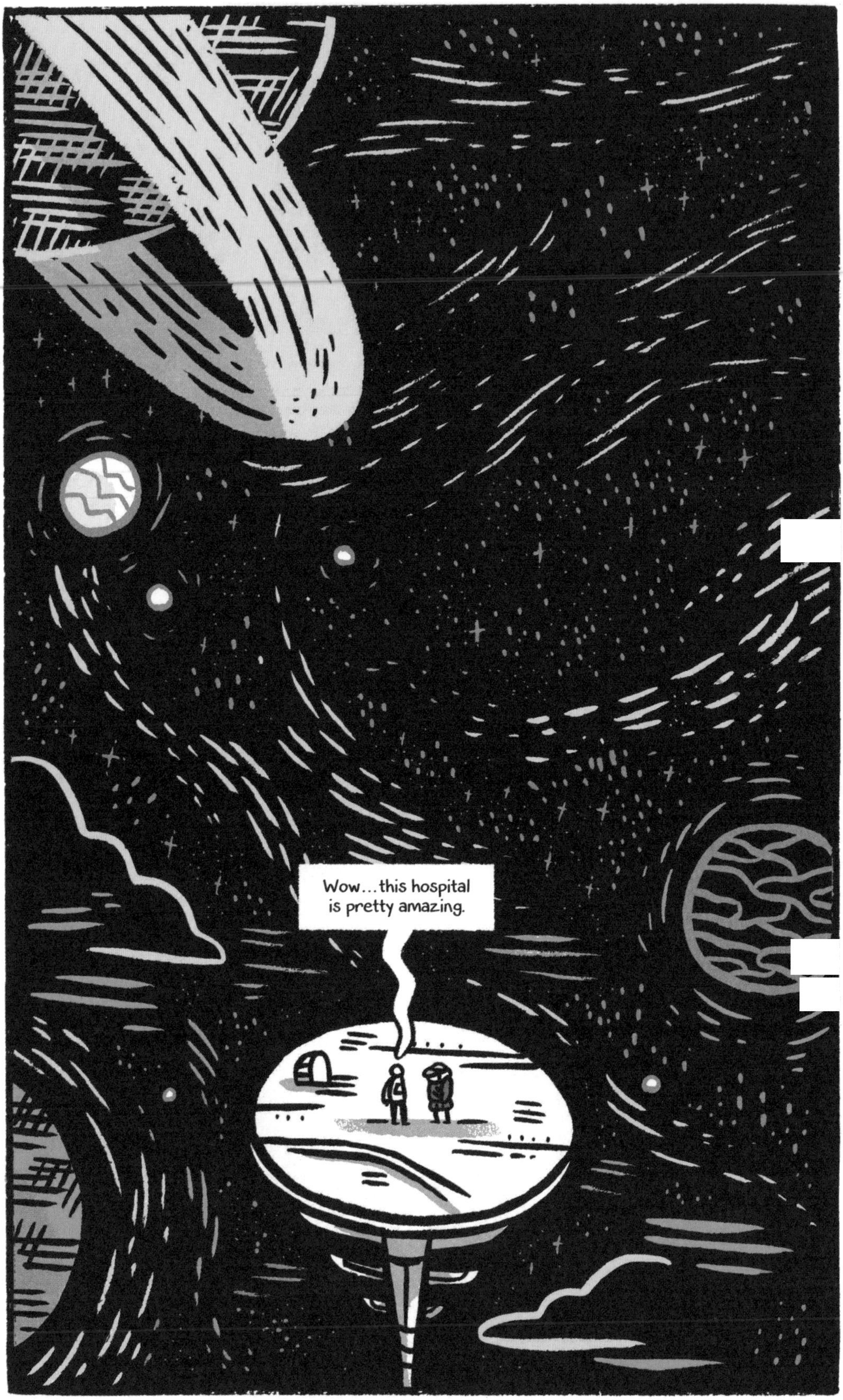
Wow... this hospital is pretty amazing.

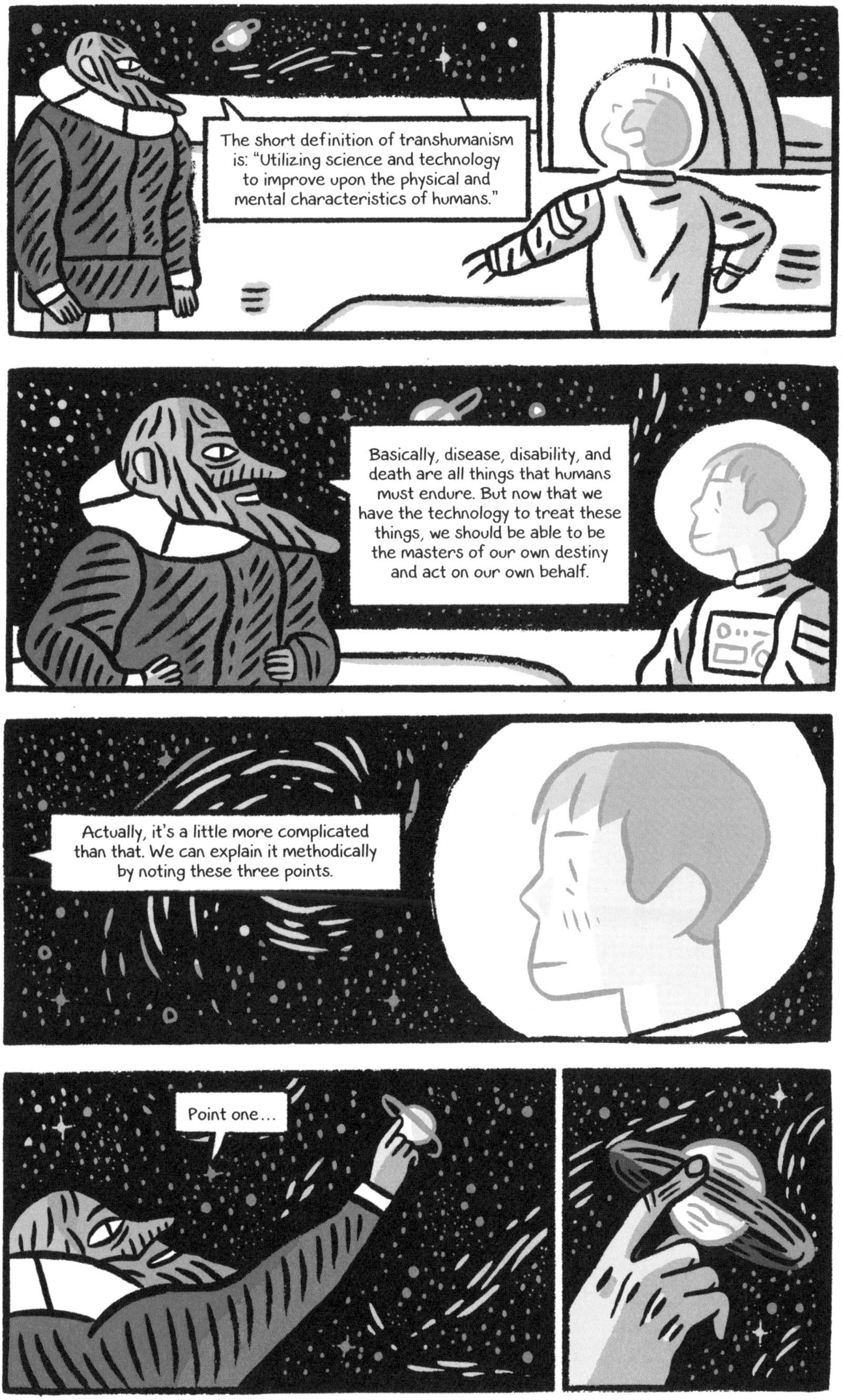
The short definition of transhumanism is: "Utilizing science and technology to improve upon the physical and mental characteristics of humans."
Basically, disease, disability, and death are all things that humans must endure. But now that we have the technology to treat these things, we should be able to be the masters of our own destiny and act on our own behalf.
Actually, it's a little more complicated than that. We can explain it methodically by noting these three points.
Point one...

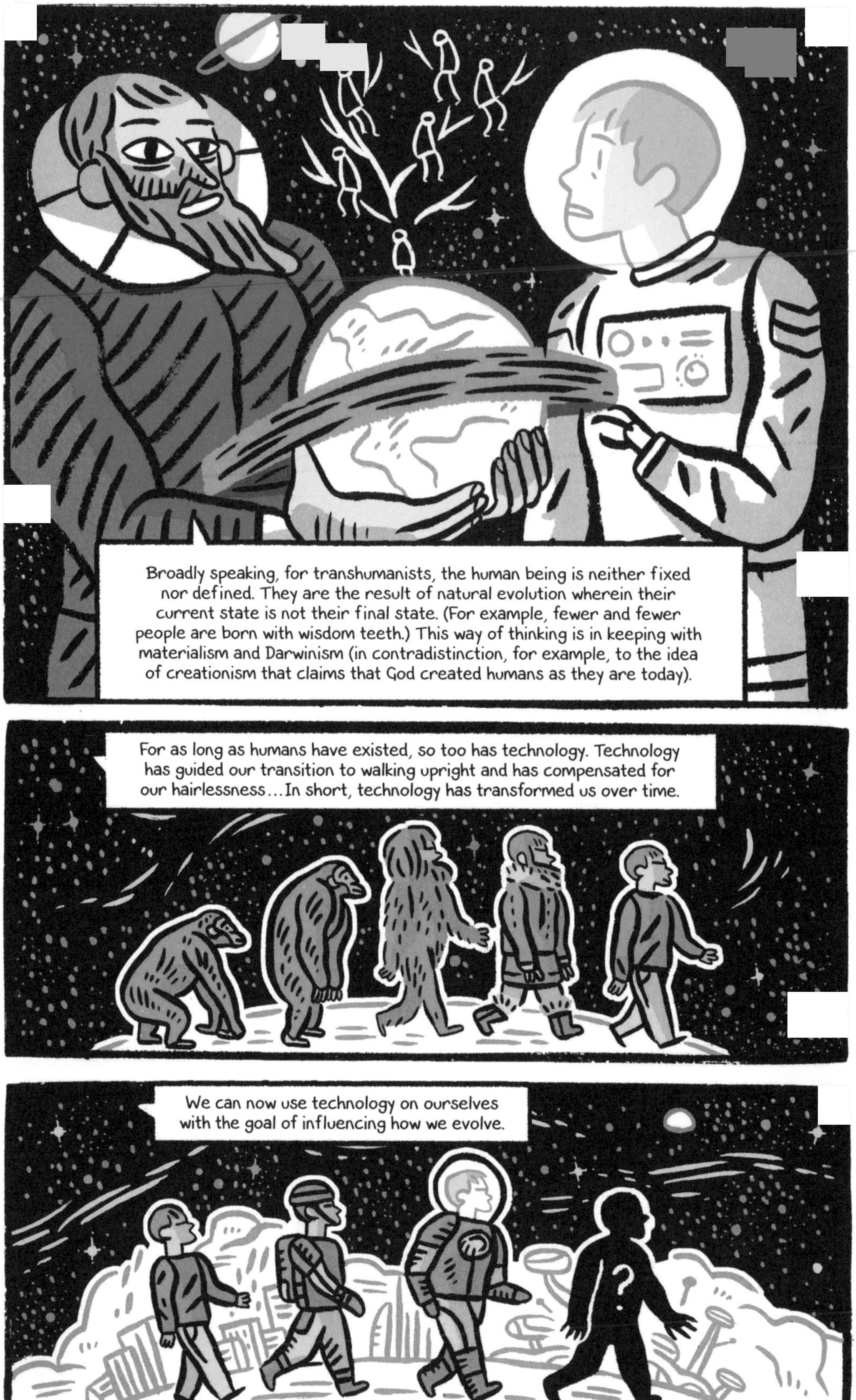
Broadly speaking, for transhumanists, the human being is neither fixed nor defined. They are the result of natural evolution wherein their current state is not their final state. (For example, fewer and fewer people are born with wisdom teeth.) This way of thinking is in keeping with materialism and Darwinism (in contradistinction, for example, to the idea of creationism that claims that God created humans as they are today).
For as long as humans have existed, so too has technology. Technology has guided our transition to walking upright and has compensated for our hairlessness…In short, technology has transformed us over time.
We can now use technology on ourselves with the goal of influencing how we evolve.

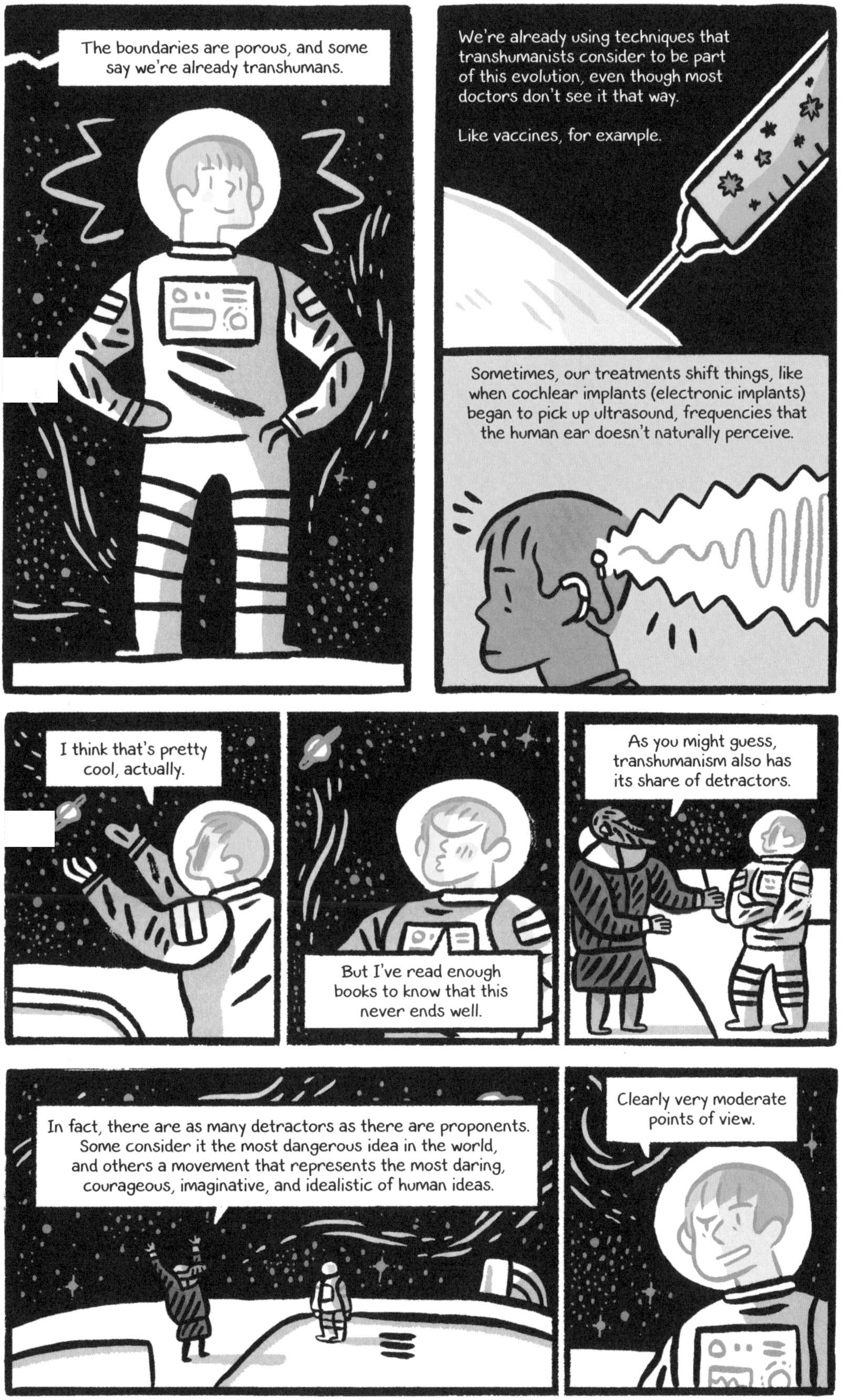
The boundaries are porous, and some say we're already transhumans.
We're already using techniques that transhumanists consider to be part of this evolution, even though most doctors don't see it that way.
Like vaccines, for example.
Sometimes, our treatments shift things, like when cochlear implants (electronic implants) began to pick up ultrasound, frequencies that the human ear doesn't naturally perceive.
I think that's pretty cool, actually.
But I've read enough books to know that this never ends well.
As you might guess, transhumanism also has its share of detractors.
In fact, there are as many detractors as there are proponents. Some consider it the most dangerous idea in the world, and others a movement that represents the most daring, courageous, imaginative, and idealistic of human ideas.
Clearly very moderate points of view.

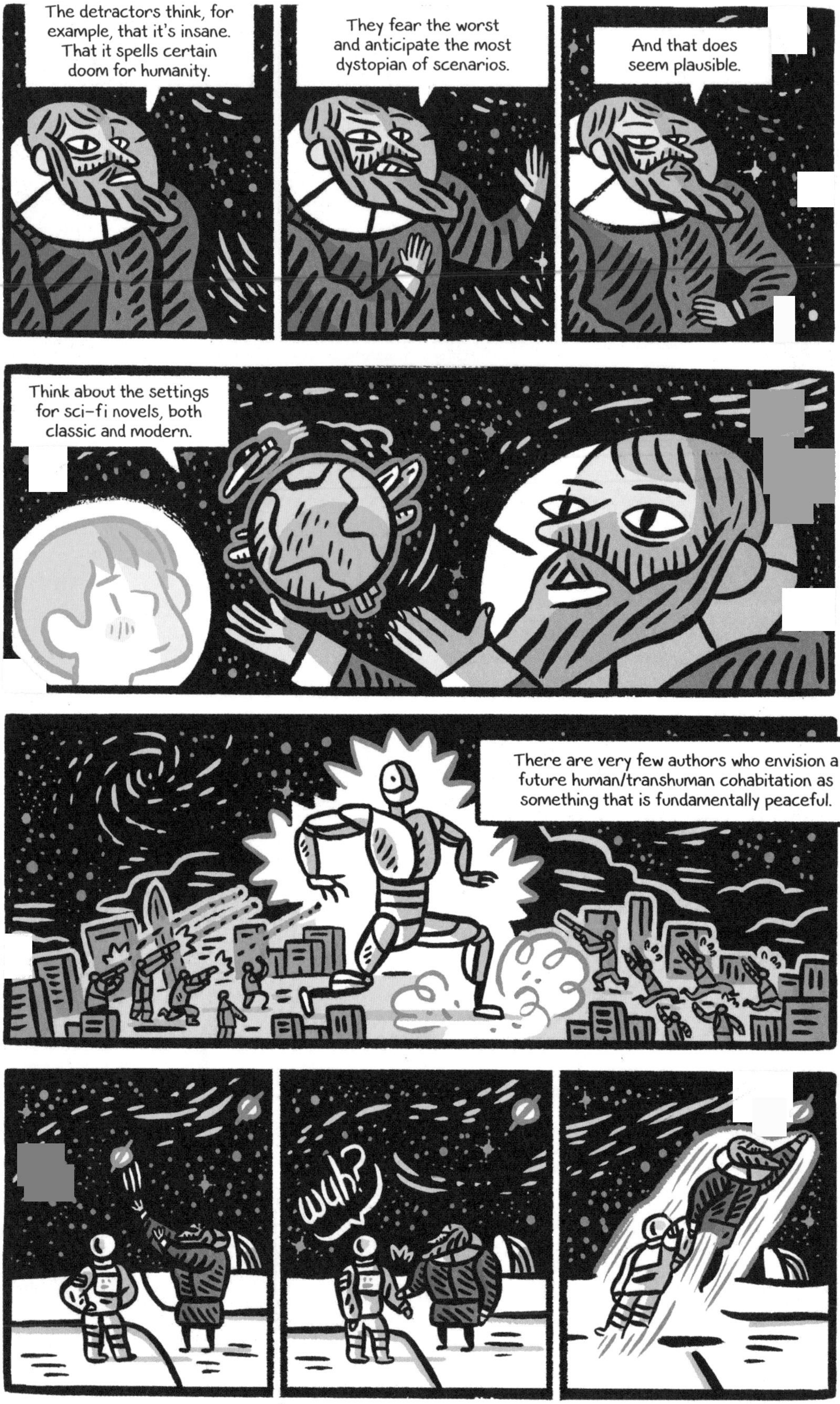

The detractors think, for example, that it's insane. That it spells certain doom for humanity.
They fear the worst and anticipate the most dystopian of scenarios.
And that does seem plausible.
Think about the settings for sci-fi novels, both classic and modern.
There are very few authors who envision a future human/transhuman cohabitation as something that is fundamentally peaceful.
wuh?

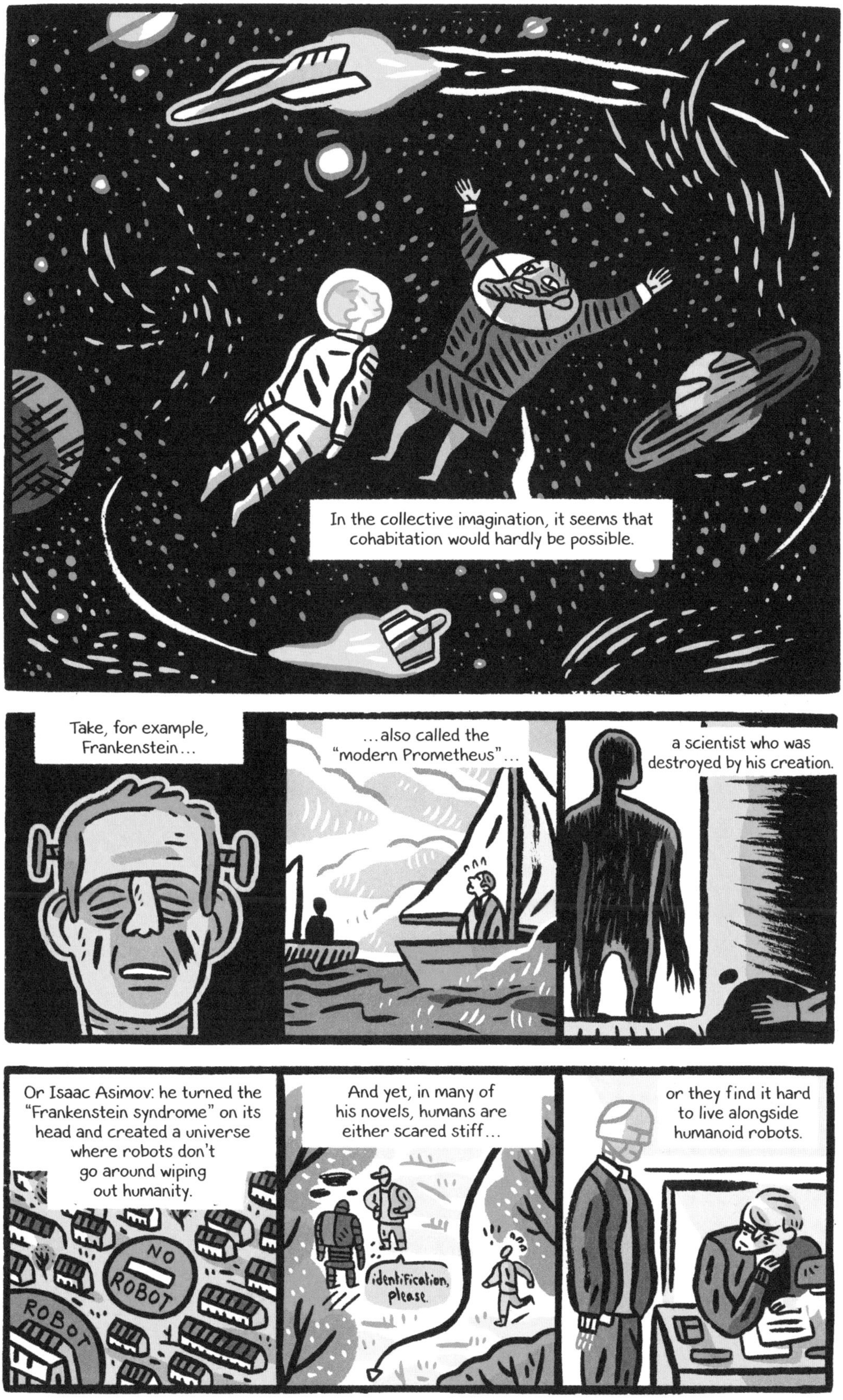
In the collective imagination, it seems that cohabitation would hardly be possible.
Take, for example, Frankenstein…
…also called the "modern Prometheus"…
a scientist who was destroyed by his creation.
Or Isaac Asimov: he turned the "Frankenstein syndrome" on its head and created a universe where robots don't go around wiping out humanity.
NO ROBOT
ROBOT
And yet, in many of his novels, humans are either scared stiff…
identification, please.
or they find it hard to live alongside humanoid robots.

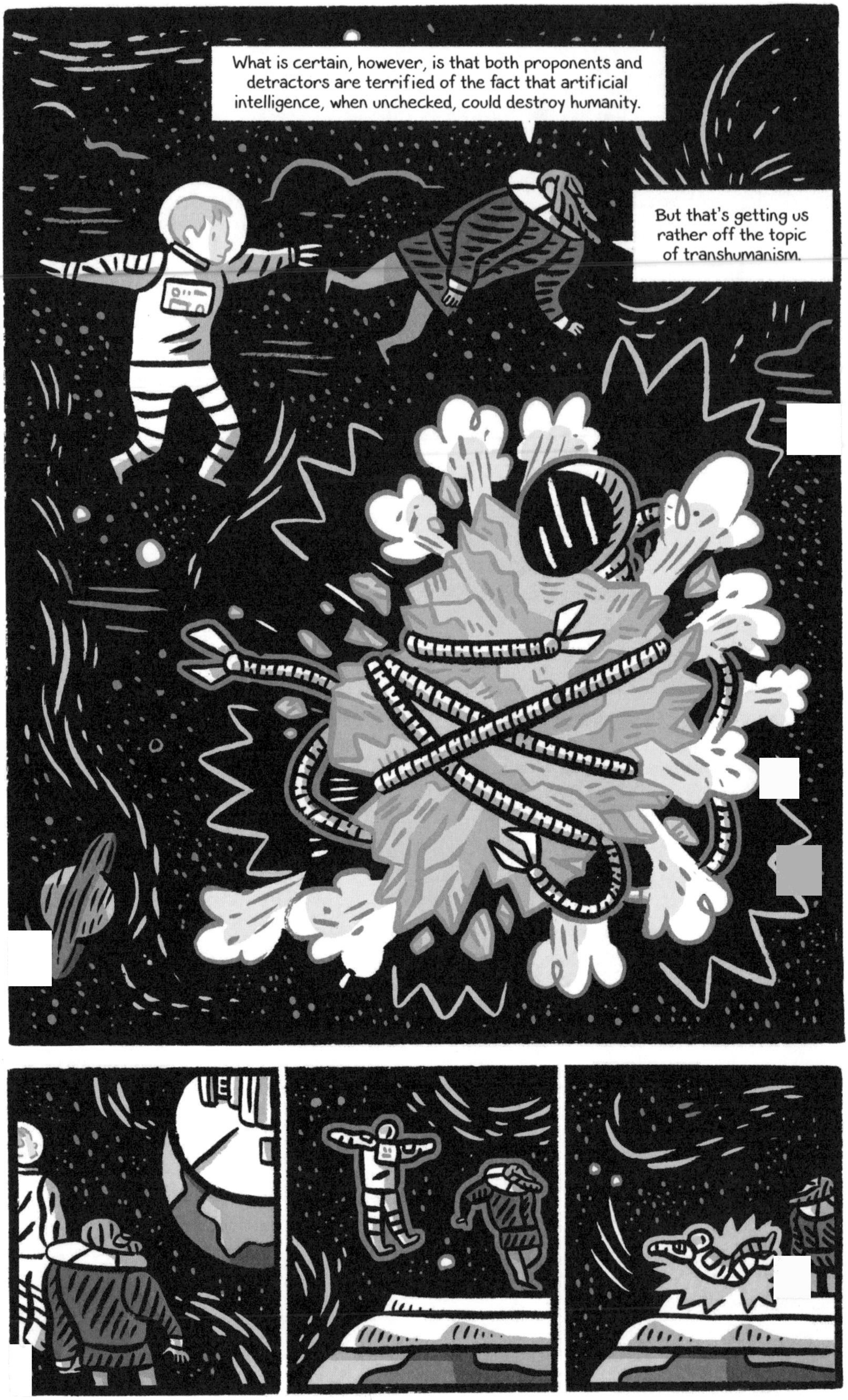
What is certain, however, is that both proponents and detractors are terrified of the fact that artificial intelligence, when unchecked, could destroy humanity.
But that's getting us rather off the topic of transhumanism.

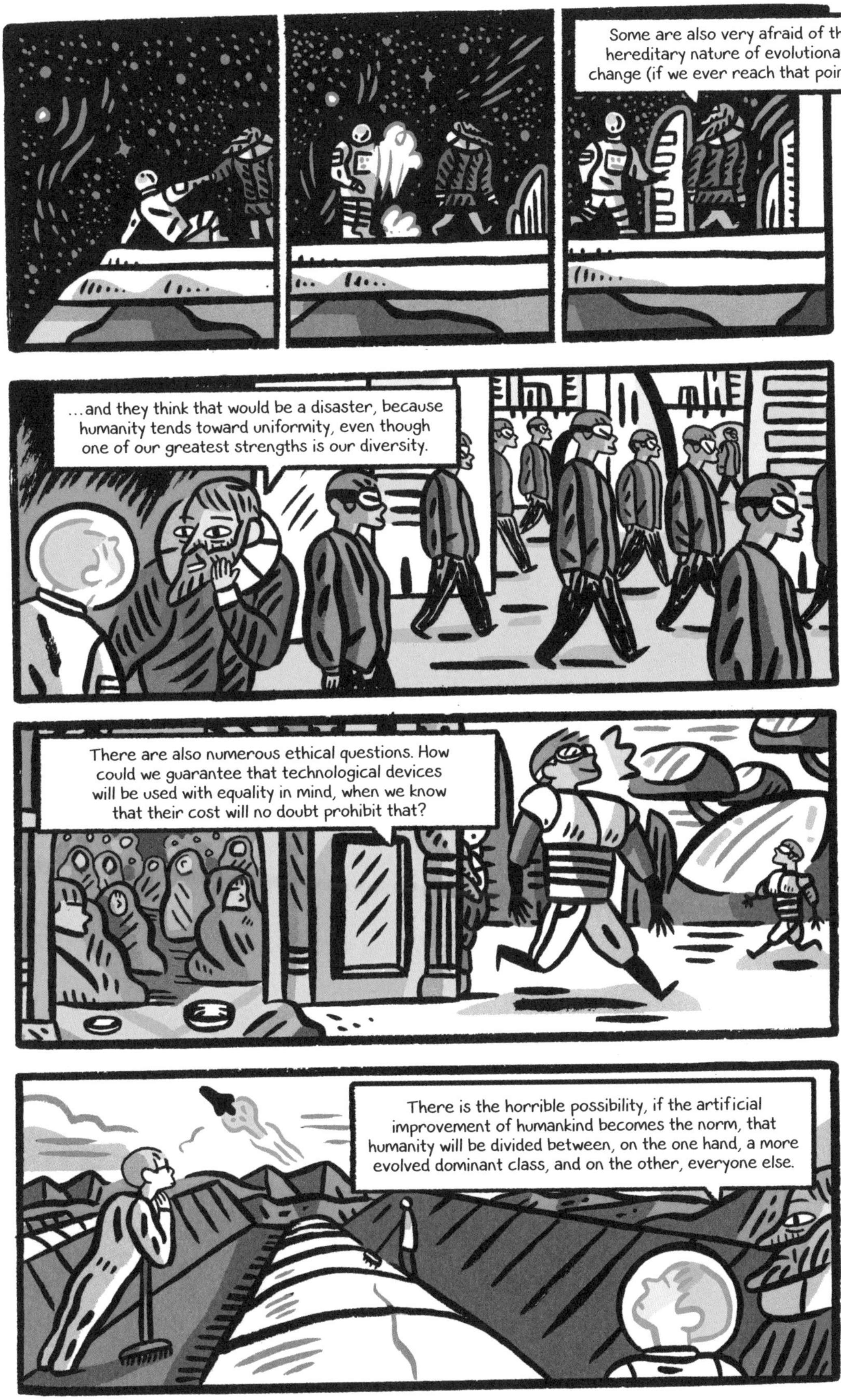

Some are also very afraid of the hereditary nature of evolutionary change (if we ever reach that point)...
...and they think that would be a disaster, because humanity tends toward uniformity, even though one of our greatest strengths is our diversity.
There are also numerous ethical questions. How could we guarantee that technological devices will be used with equality in mind, when we know that their cost will no doubt prohibit that?
There is the horrible possibility, if the artificial improvement of humankind becomes the norm, that humanity will be divided between, on the one hand, a more evolved dominant class, and on the other, everyone else.

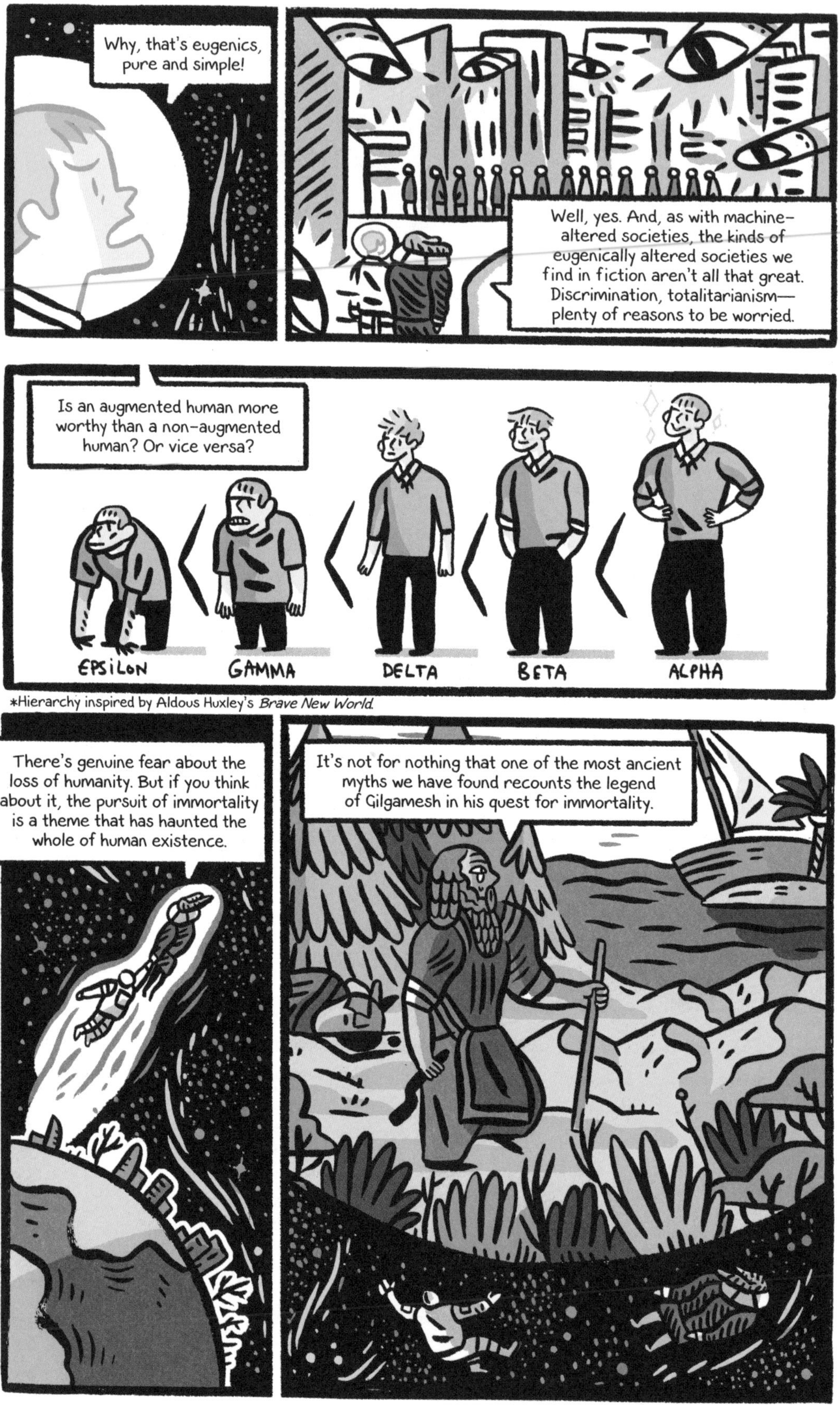
Why, that's eugenics, pure and simple!
Well, yes. And, as with machine-altered societies, the kinds of eugenically altered societies we find in fiction aren't all that great. Discrimination, totalitarianism—plenty of reasons to be worried.
Is an augmented human more worthy than a non-augmented human? Or vice versa?
EPSILON
GAMMA
DELTA
BETA
ALPHA
*Hierarchy inspired by Aldous Huxley's *Brave New World*
There's genuine fear about the loss of humanity. But if you think about it, the pursuit of immortality is a theme that has haunted the whole of human existence.
It's not for nothing that one of the most ancient myths we have found recounts the legend of Gilgamesh in his quest for immortality.

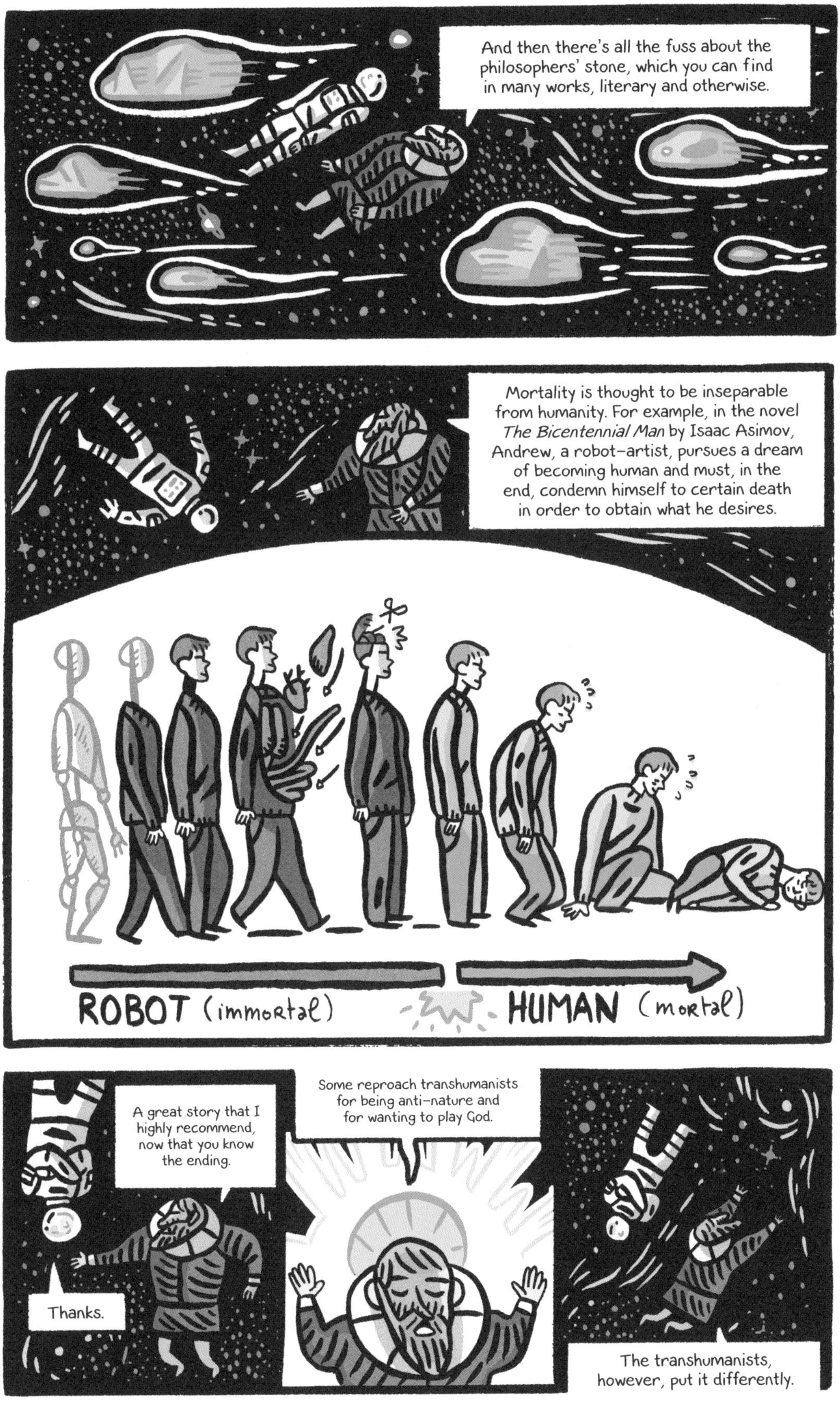

And then there's all the fuss about the philosophers' stone, which you can find in many works, literary and otherwise.
Mortality is thought to be inseparable from humanity. For example, in the novel *The Bicentennial Man* by Isaac Asimov, Andrew, a robot-artist, pursues a dream of becoming human and must, in the end, condemn himself to certain death in order to obtain what he desires.
ROBOT (immortal)
HUMAN (mortal)
A great story that I highly recommend, now that you know the ending.
Thanks.
Some reproach transhumanists for being anti-nature and for wanting to play God.
The transhumanists, however, put it differently.

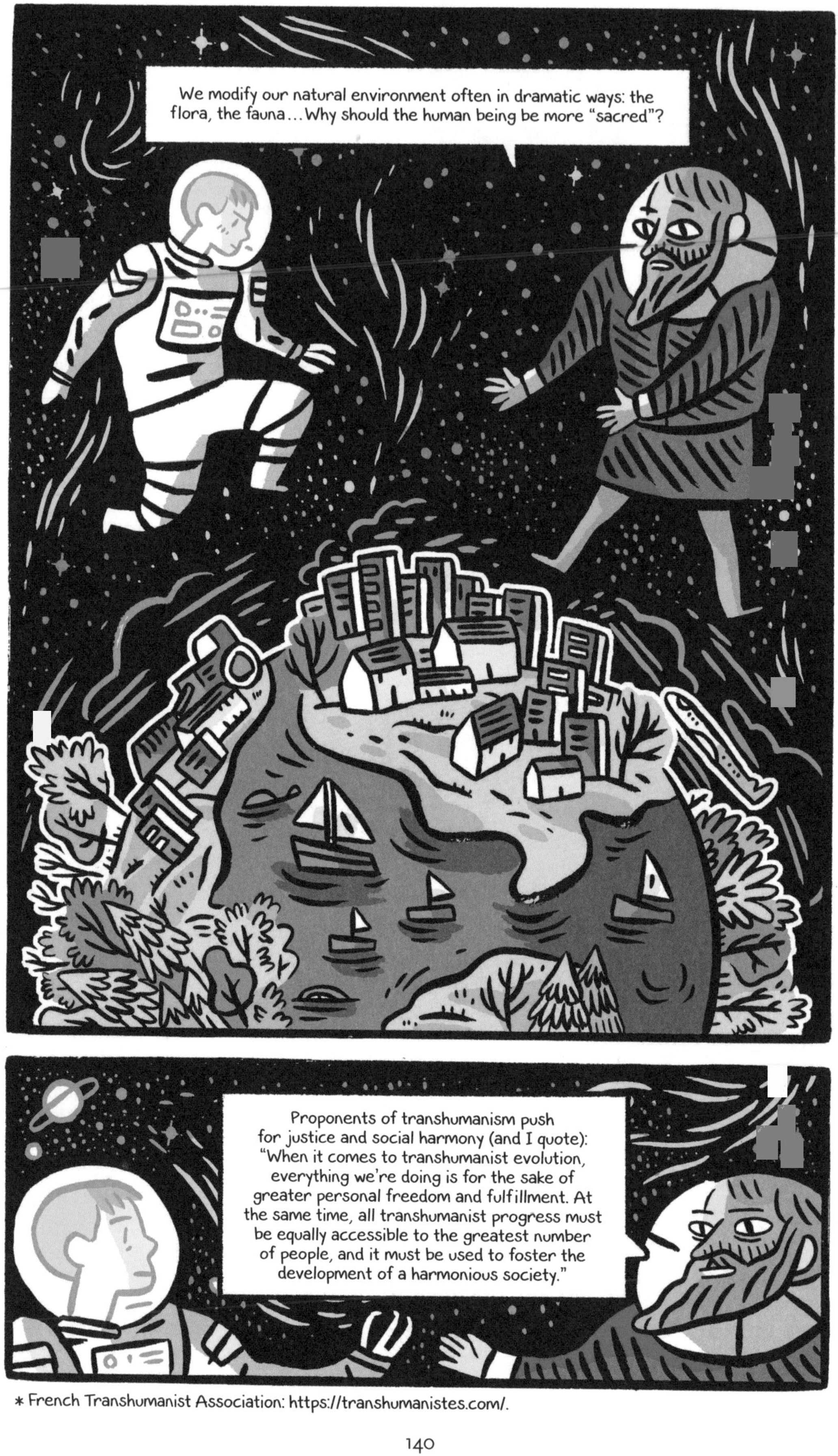
We modify our natural environment often in dramatic ways: the flora, the fauna... Why should the human being be more "sacred"?
Proponents of transhumanism push for justice and social harmony (and I quote): "When it comes to transhumanist evolution, everything we're doing is for the sake of greater personal freedom and fulfillment. At the same time, all transhumanist progress must be equally accessible to the greatest number of people, and it must be used to foster the development of a harmonious society."

* French Transhumanist Association: https://transhumanistes.com/.

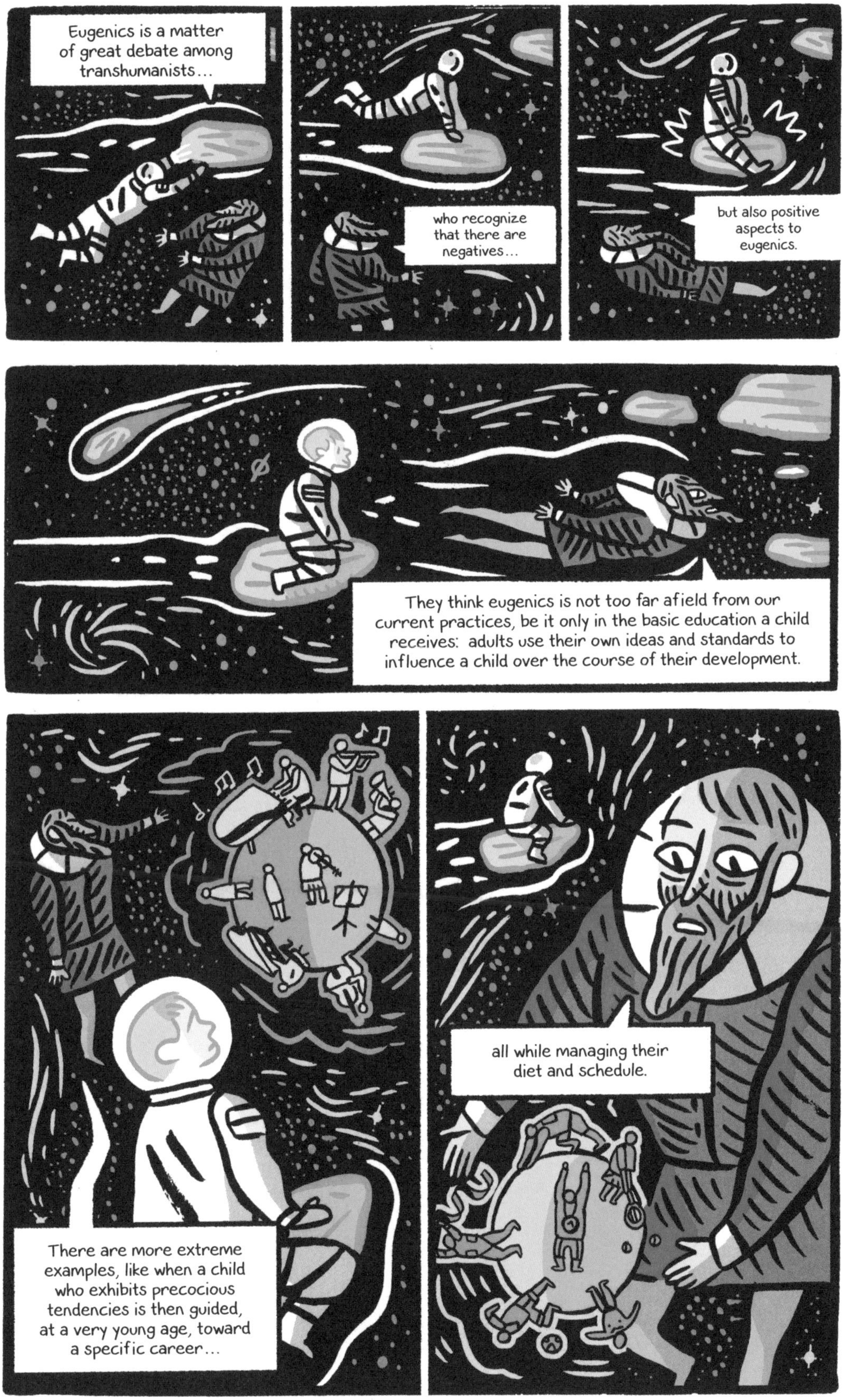
Eugenics is a matter of great debate among transhumanists...
who recognize that there are negatives...
but also positive aspects to eugenics.
They think eugenics is not too far afield from our current practices, be it only in the basic education a child receives: adults use their own ideas and standards to influence a child over the course of their development.
There are more extreme examples, like when a child who exhibits precocious tendencies is then guided, at a very young age, toward a specific career...
all while managing their diet and schedule.

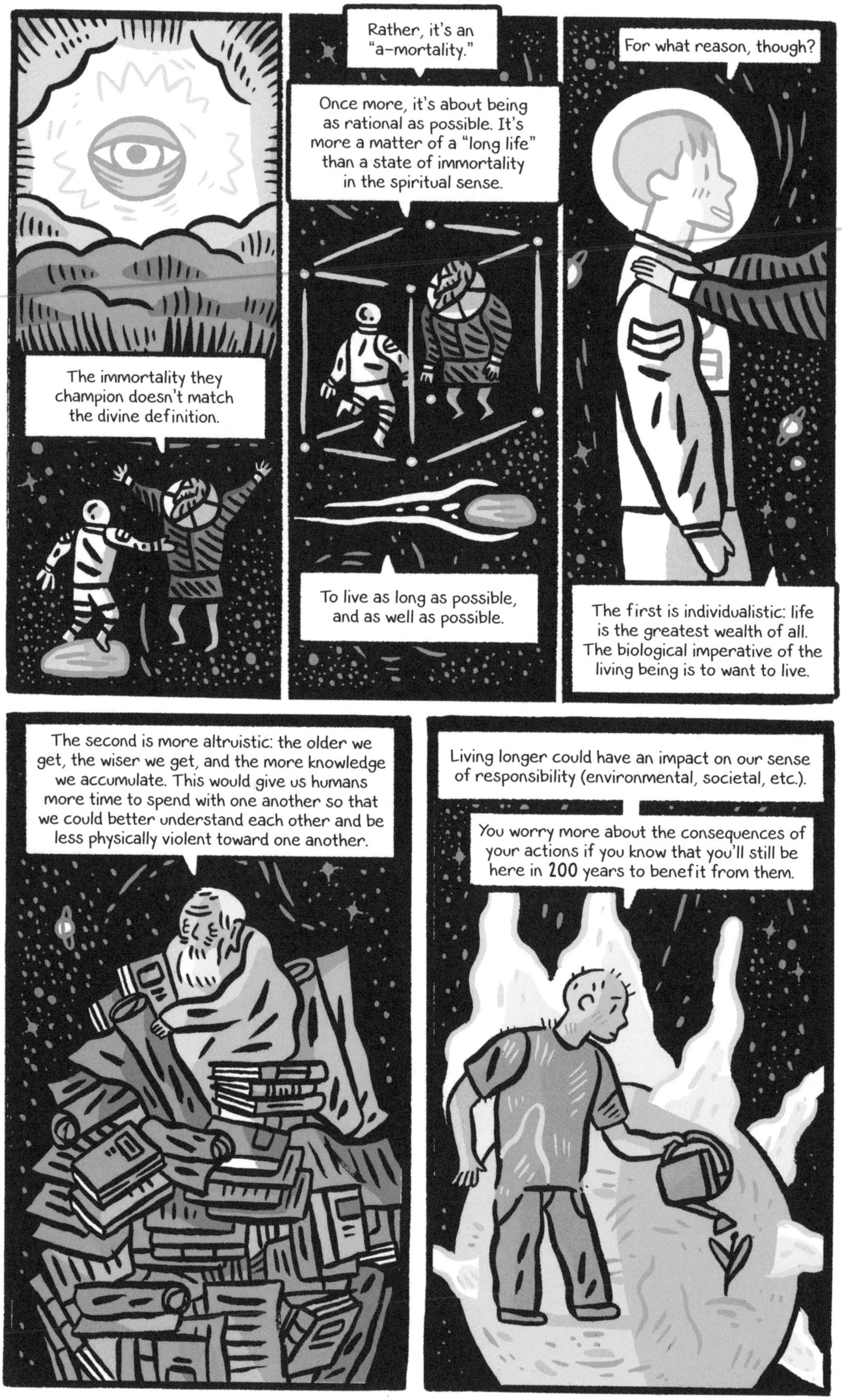
The immortality they champion doesn't match the divine definition.
Rather, it's an "a-mortality."
Once more, it's about being as rational as possible. It's more a matter of a "long life" than a state of immortality in the spiritual sense.
To live as long as possible, and as well as possible.
For what reason, though?
The first is individualistic: life is the greatest wealth of all. The biological imperative of the living being is to want to live.
The second is more altruistic: the older we get, the wiser we get, and the more knowledge we accumulate. This would give us humans more time to spend with one another so that we could better understand each other and be less physically violent toward one another.
Living longer could have an impact on our sense of responsibility (environmental, societal, etc.).
You worry more about the consequences of your actions if you know that you'll still be here in 200 years to benefit from them.

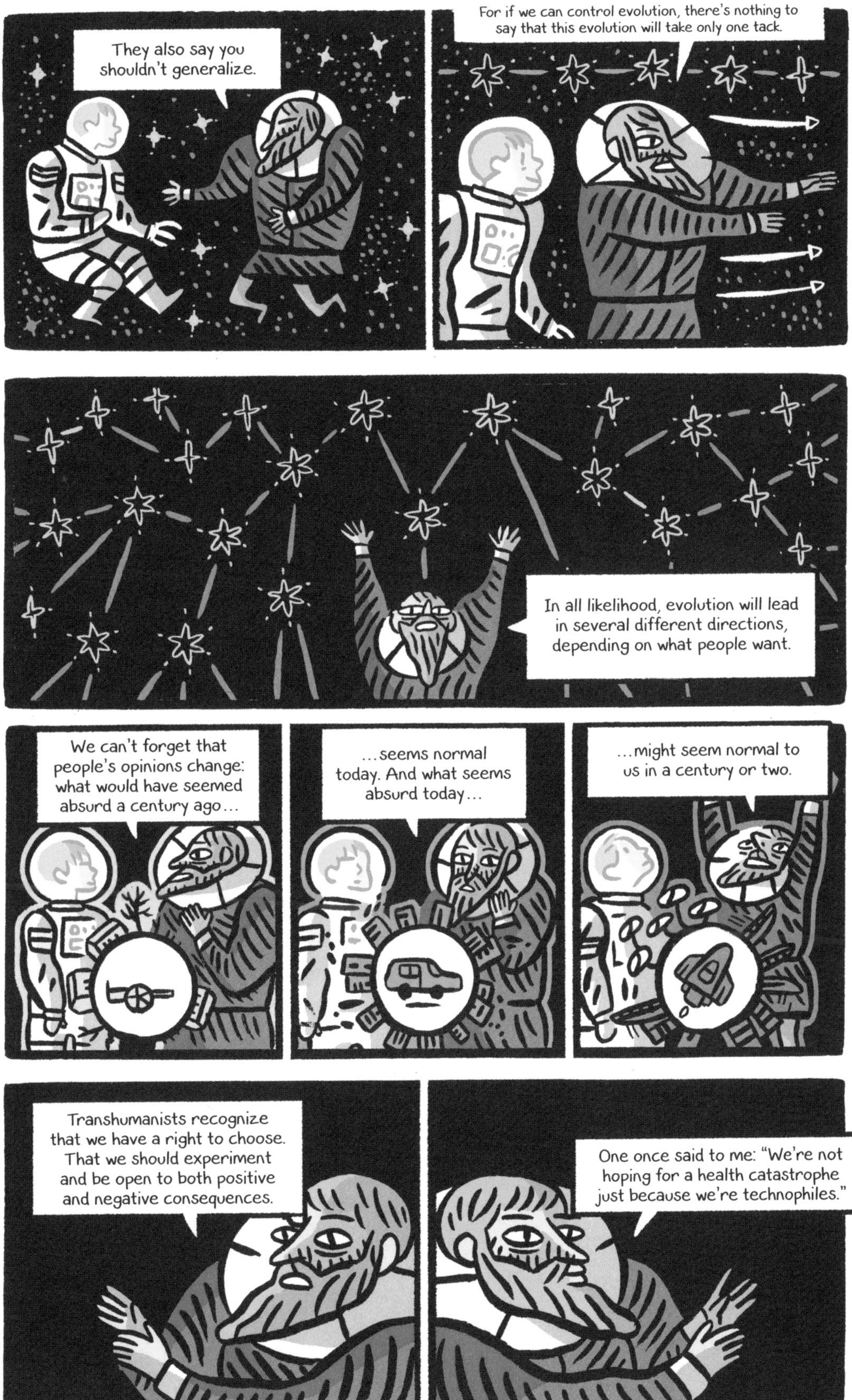
They also say you shouldn't generalize.
For if we can control evolution, there's nothing to say that this evolution will take only one tack.
In all likelihood, evolution will lead in several different directions, depending on what people want.
We can't forget that people's opinions change: what would have seemed absurd a century ago…
…seems normal today. And what seems absurd today…
…might seem normal to us in a century or two.
Transhumanists recognize that we have a right to choose. That we should experiment and be open to both positive and negative consequences.
One once said to me: "We're not hoping for a health catastrophe just because we're technophiles."

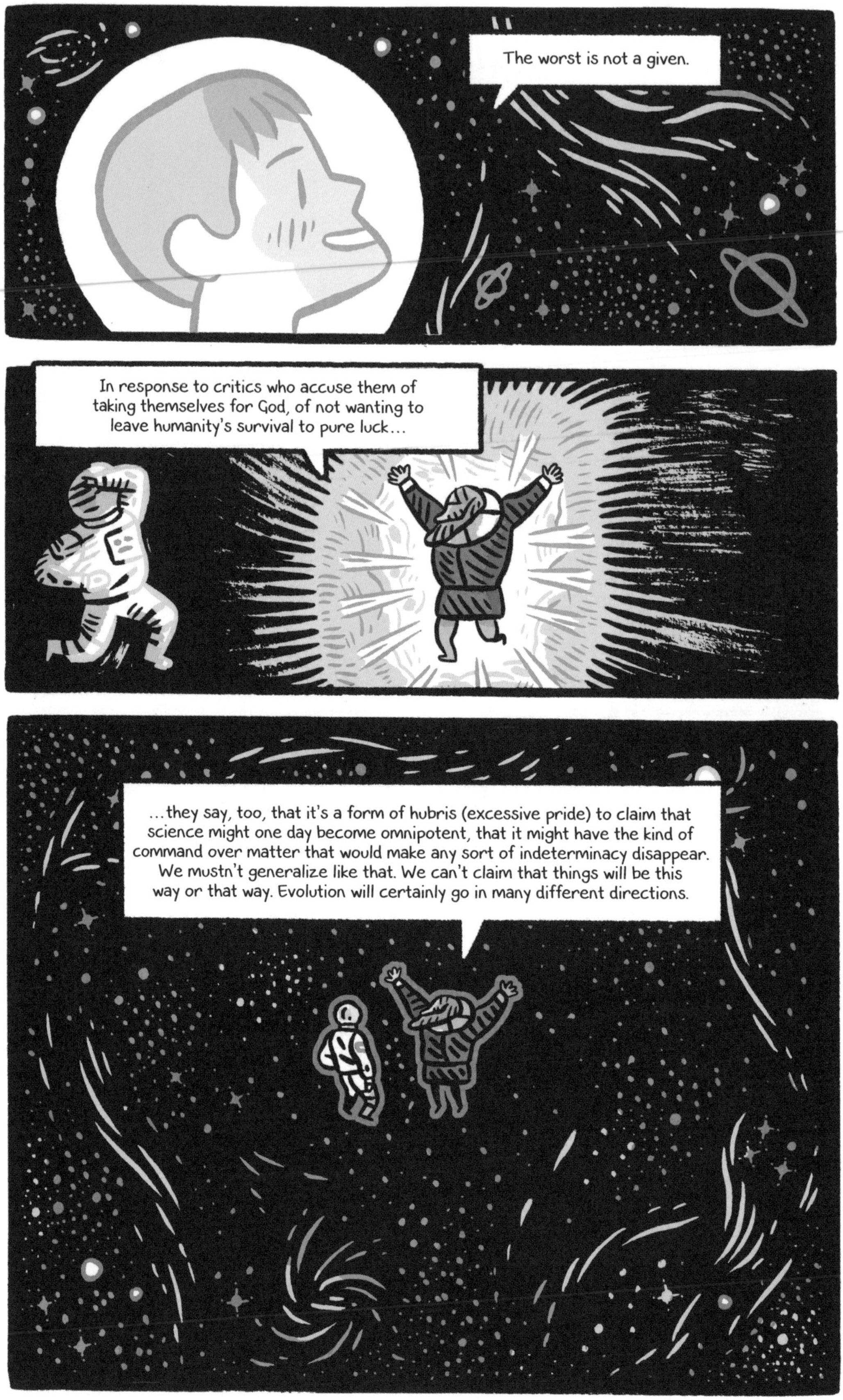
The worst is not a given.
In response to critics who accuse them of taking themselves for God, of not wanting to leave humanity's survival to pure luck...
...they say, too, that it's a form of hubris (excessive pride) to claim that science might one day become omnipotent, that it might have the kind of command over matter that would make any sort of indeterminacy disappear. We mustn't generalize like that. We can't claim that things will be this way or that way. Evolution will certainly go in many different directions.

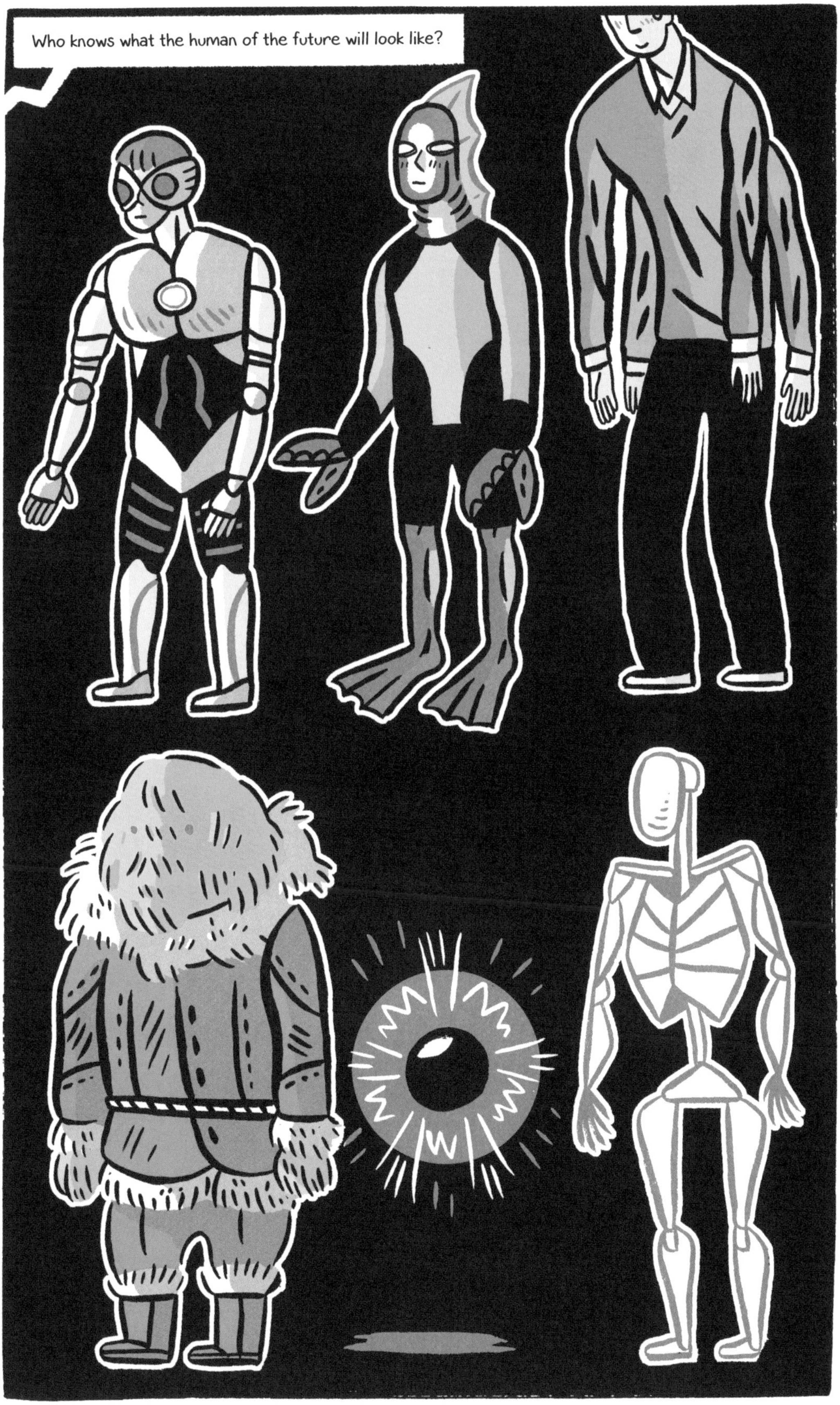
Who knows what the human of the future will look like?

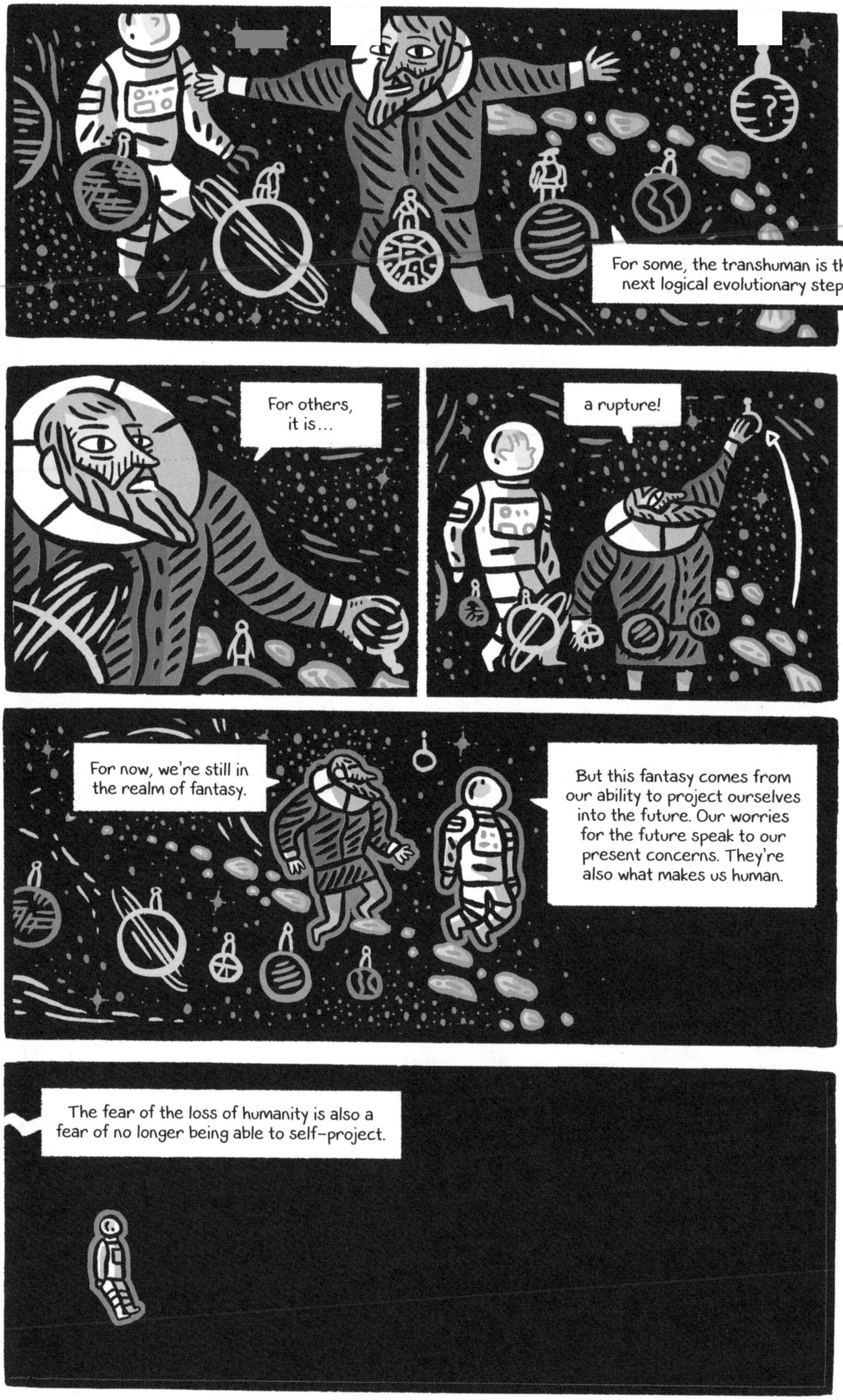
For some, the transhuman is the next logical evolutionary step.
For others, it is...
a rupture!
For now, we're still in the realm of fantasy.
But this fantasy comes from our ability to project ourselves into the future. Our worries for the future speak to our present concerns. They're also what makes us human.
The fear of the loss of humanity is also a fear of no longer being able to self-project.

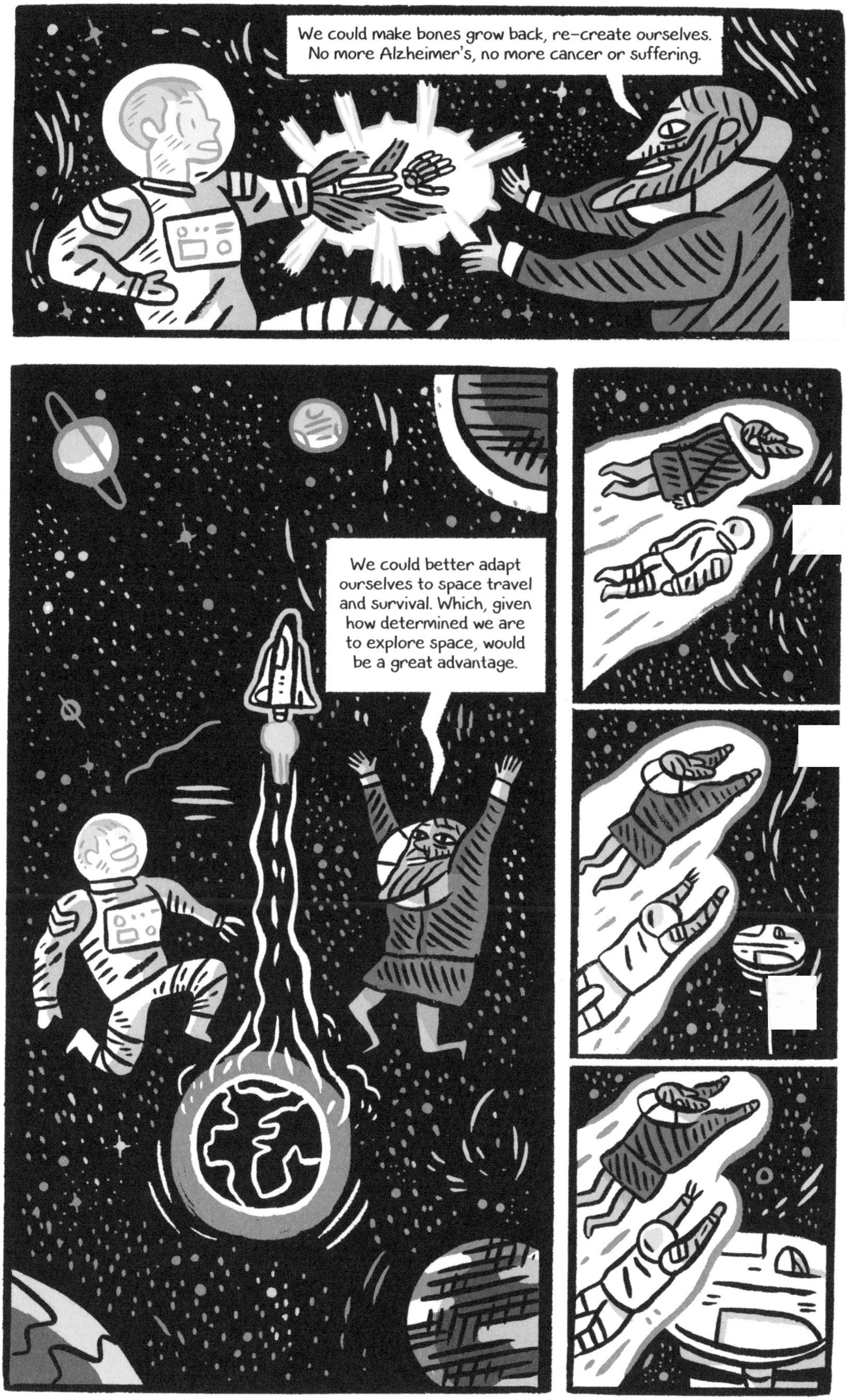
We could make bones grow back, re-create ourselves. No more Alzheimer's, no more cancer or suffering.
We could better adapt ourselves to space travel and survival. Which, given how determined we are to explore space, would be a great advantage.

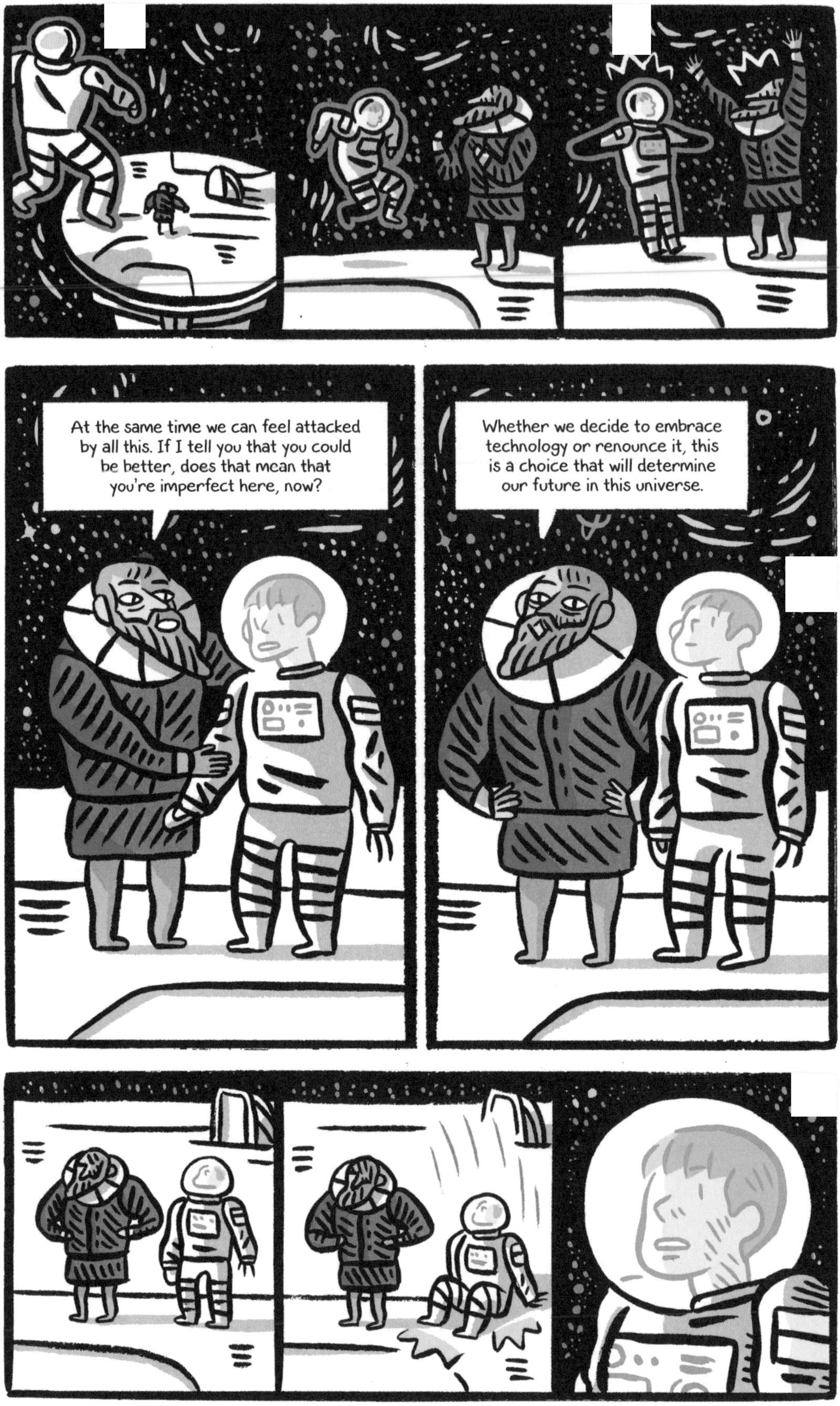
At the same time we can feel attacked by all this. If I tell you that you could be better, does that mean that you're imperfect here, now?
Whether we decide to embrace technology or renounce it, this is a choice that will determine our future in this universe.

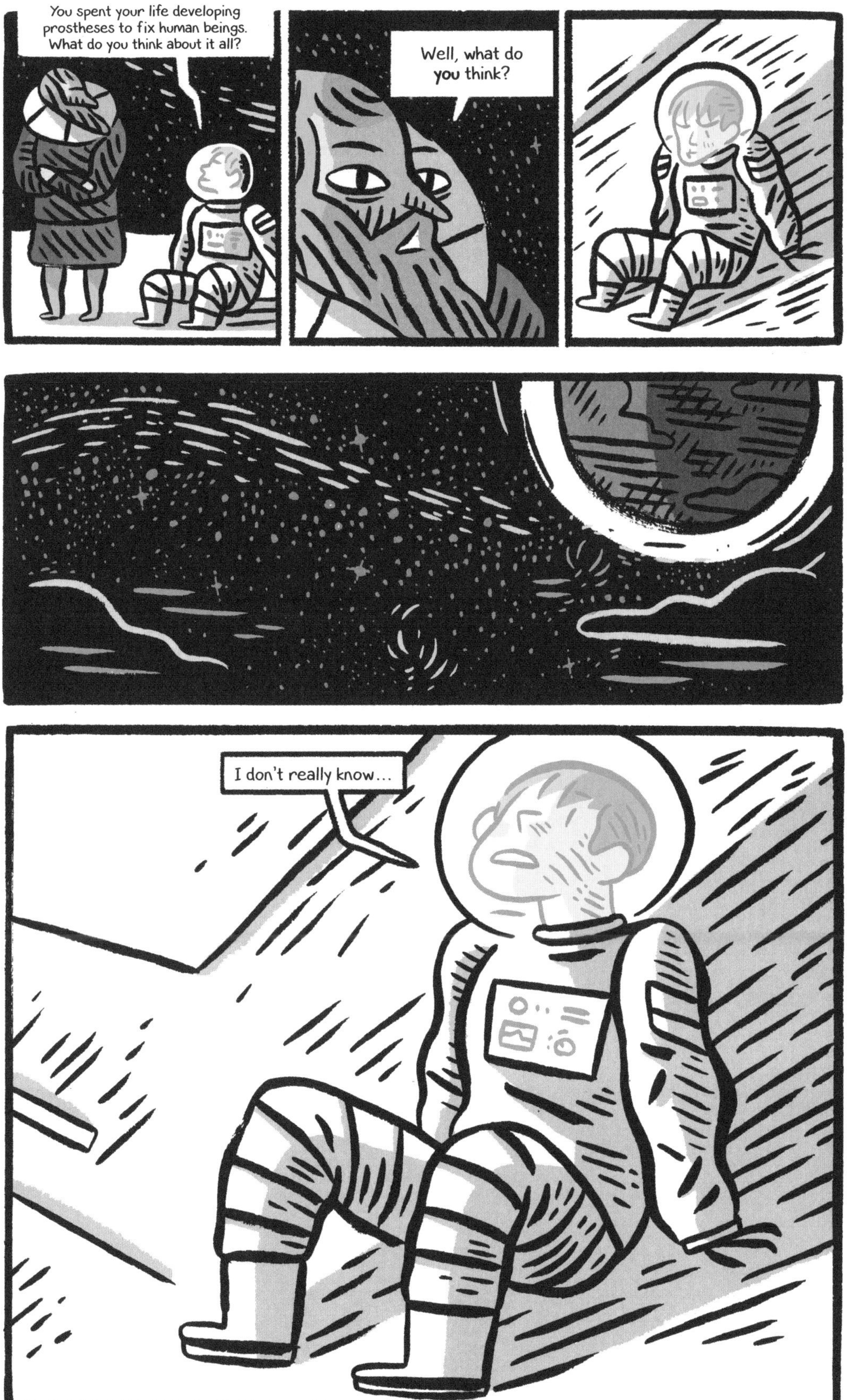
You spent your life developing prostheses to fix human beings. What do you think about it all?
Well, what do you think?
I don't really know...

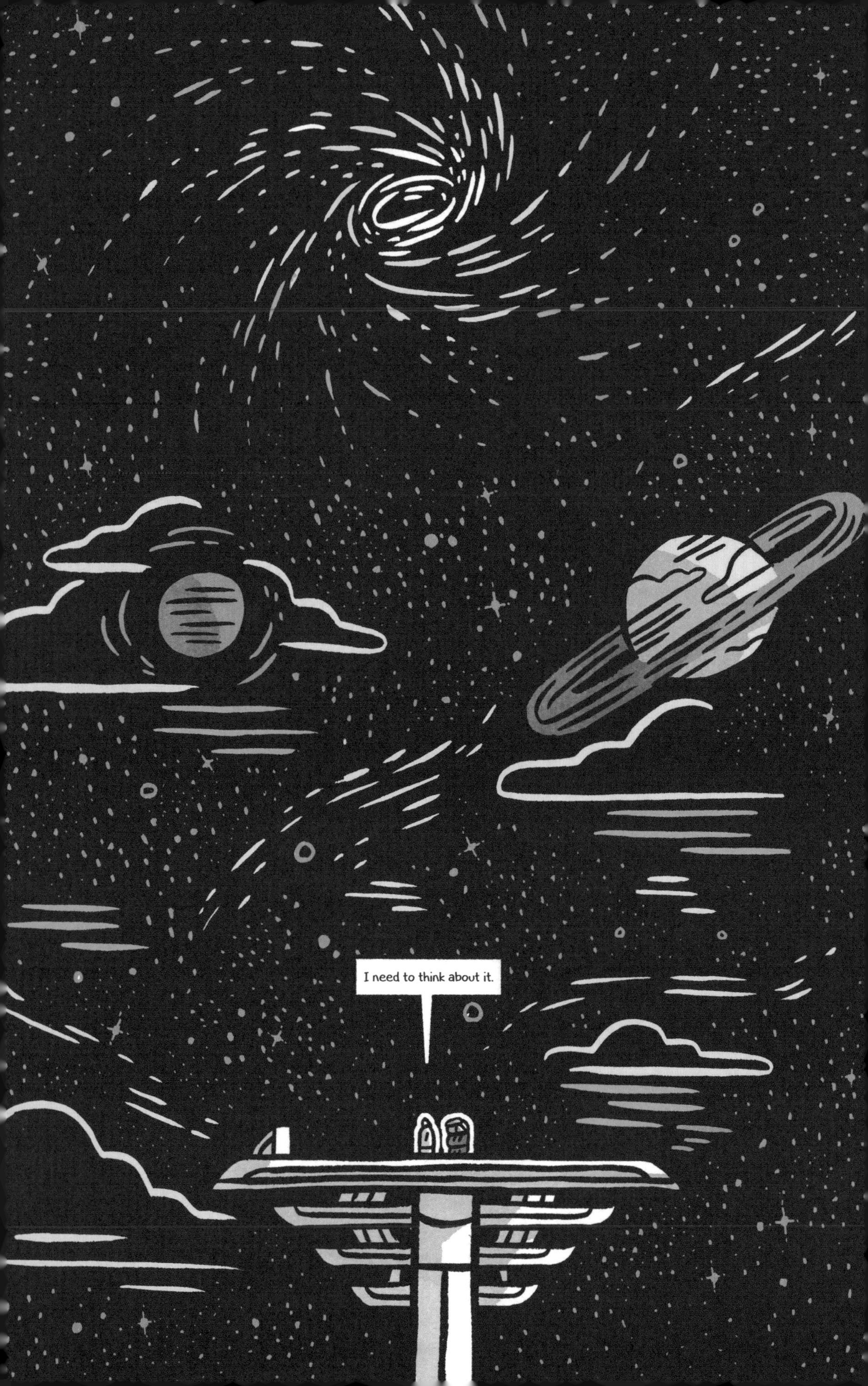
I need to think about it.

EPILOGUE

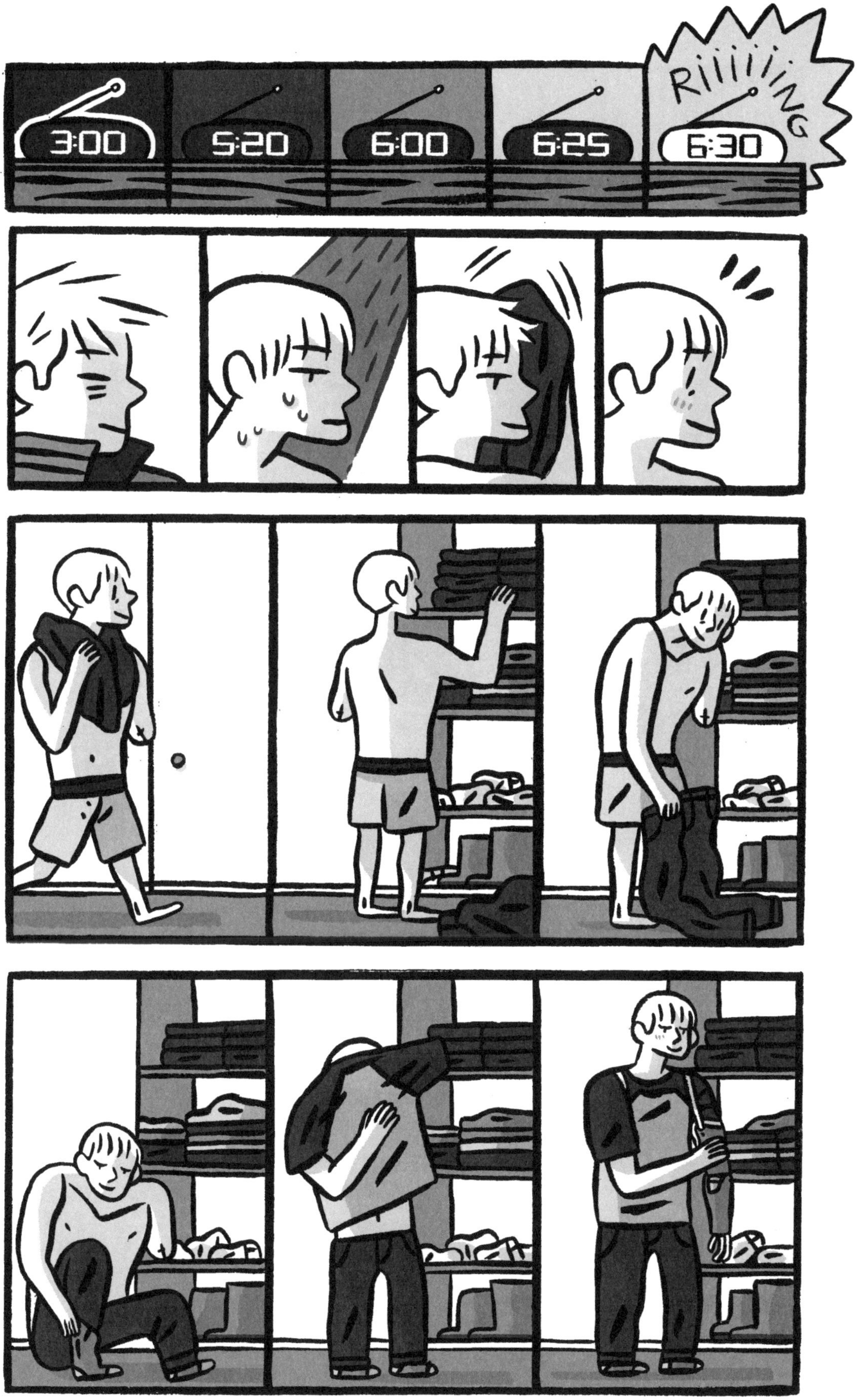
3:00
5:20
6:00
6:25
6:30
RIIIIING

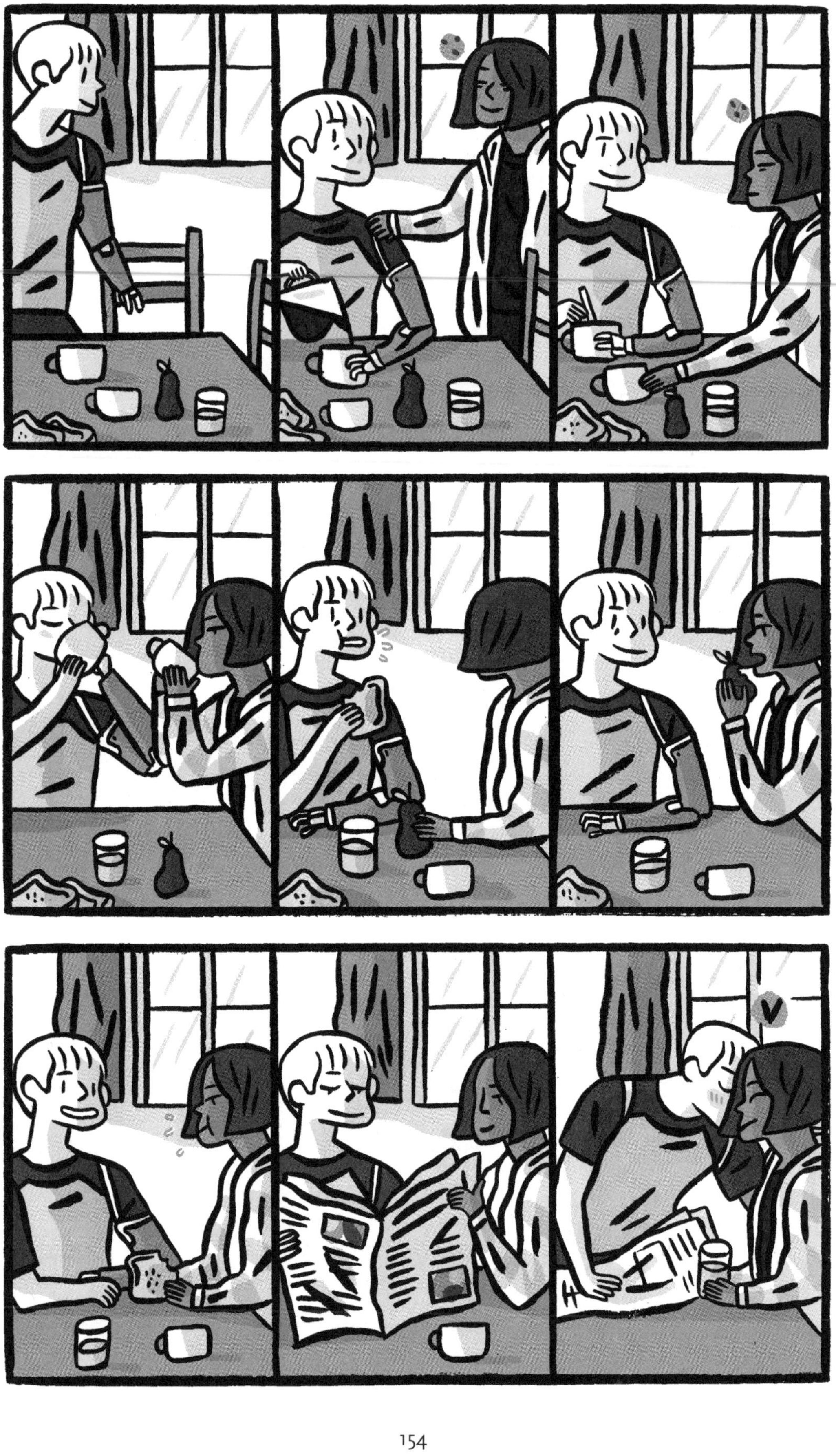

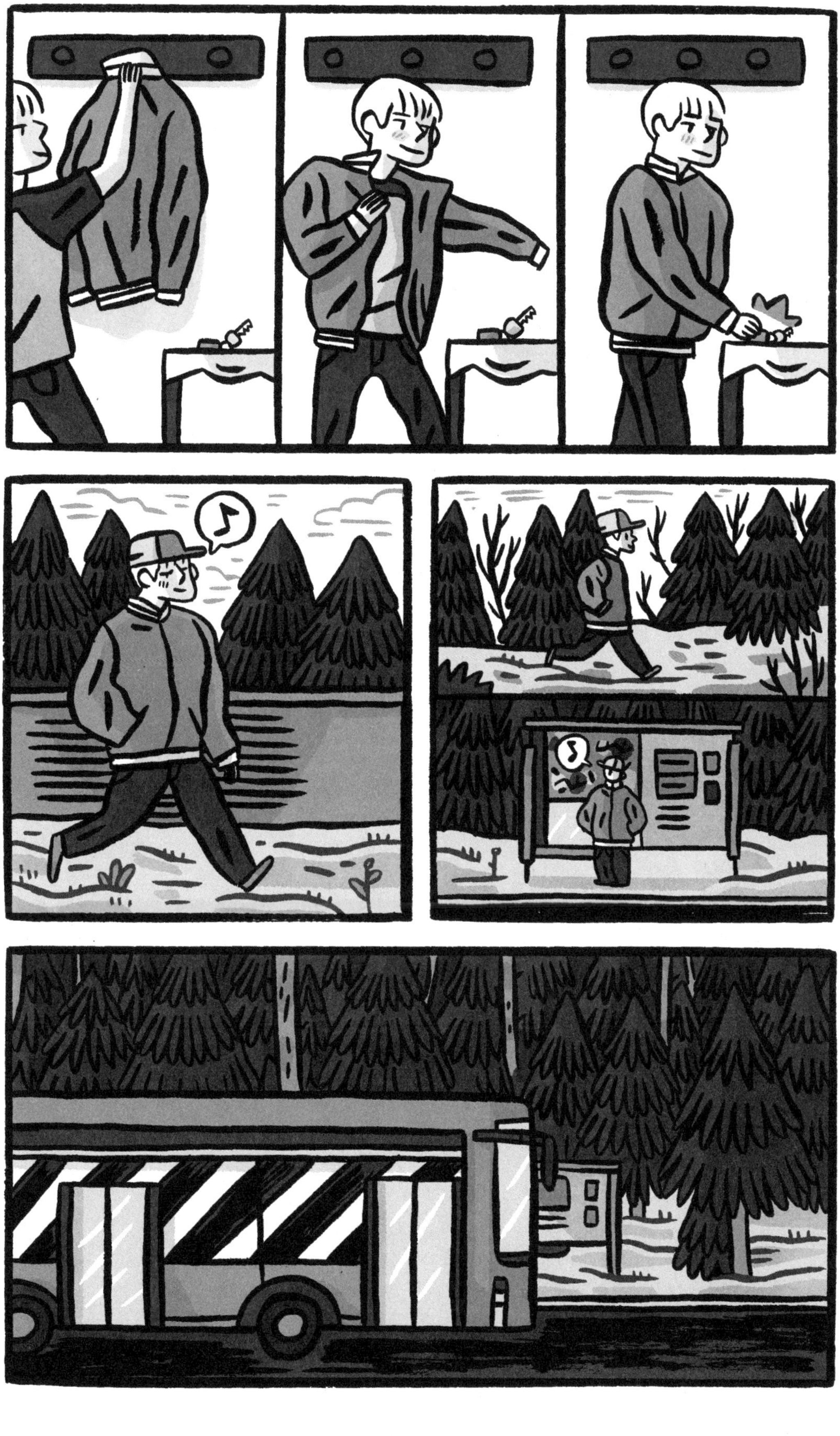

*A huge thank you to Dr. Dominique Hasbourn for sharing both his time and his knowledge.
My thanks also to the AFT (Association Française Transhumaniste) as well as to members of the ADEPA (Association de Défense et d'Étude des Personnes Amputées) and to Fabrice Sabre for agreeing to meet with me and answer my questions. I hope I have managed to accurately convey the content of our interviews.*

Thank you to Marion and Boulet for staying with me over the course of this project.

Thank you to Guillaume, Victoria, Jacque, and Blandine for cooking food for me and keeping me company; to Mathilde, Claire, and Athitaya for helping me edit the text; and to all my friends and former professors at the DIS without whom I would never have had the idea to make this book.

Finally, thanks to my family for their unwavering support.

D'Héloïse Chochois, aux éditions L'Agrume :
- *Globe-trotters* - avec Emmanuelle Mardesson

heloisechochois.tumblr.com

Library of Congress Cataloging-in-Publication Data

Names: Chochois, Héloïse, 1991– author, artist.
Title: The body factory : from the first prosthetics to the augmented human / story and art, Héloïse Chochois.
Other titles: Fabrique des corps. English
Description: University Park, Pennsylvania : The Pennsylvania State University Press, Graphic Mundi, [2021] | Translation of: La fabrique des corps: Des premières prothèses à l'humain augmenté. Première édition. Paris : Delcourt, [2017]
Summary: "A graphic novel exploring amputation, revealing details about famous amputees throughout history, the invention of the tourniquet, phantom limb syndrome, types of prostheses, and transhumanist technologies"—Provided by publisher.
Identifiers: LCCN 2021001783 | ISBN 9780271087061 (paperback ; alk. paper)
Subjects: MESH: Paré, Ambroise, 1510?–1590. | Artificial Limbs—history | Amputees—history | Amputation—methods | Amputation—history | Phantom Limb | Biomedical Enhancement | Famous Persons | Graphic Novel
Classification: LCC RD560 | NLM WE 17 | DDC 617.5/8—dc23
LC record available at https://lccn.loc.gov/2021001783

Printed in the United States of America
Published by The Pennsylvania State University Press,
University Park, PA 16802-1003

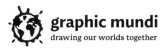

graphic mundi
drawing our worlds together

Graphic Mundi is an imprint of The Pennsylvania State University Press.

Translated by Kendra Boileau

Graphic Design: Héloïse Chochois and Trait pour Trait

Originally published as *La Fabrique des corps* by Héloïse Chochois

The Pennsylvania State University Press is a member of the Association of University Presses.

It is the policy of The Pennsylvania State University Press to use acid-free paper. Publications on uncoated stock satisfy the minimum requirements of American National Standard for Information Sciences—Permanence of Paper for Printed Library Material, ANSI Z39.48–1992.

7/22